A Mysterious Universe

A Mysterious Universe

Quantum Mechanics, Relativity, and Cosmology for Everyone

M. SUHAIL ZUBAIRY
Texas A&M University

Great Clarendon Street, Oxford, OX2 6DP,
United Kingdom

Oxford University Press is a department of the University of Oxford.
It furthers the University's objective of excellence in research, scholarship,
and education by publishing worldwide. Oxford is a registered trade mark of
Oxford University Press in the UK and in certain other countries

Published in the United States of America by Oxford University Press
198 Madison Avenue, New York, NY 10016, United States of America

British Library Cataloguing in Publication Data
Data available

Library of Congress Control Number: 2023905809

ISBN 978–0–19–888306–7

DOI: 10.1093/oso/9780198883067.001.0001

Printed and bound by
CPI Group (UK) Ltd, Croydon, CR0 4YY

Dedicated to

Marlan O. Scully

Mentor, colleague, and friend for life

Foreword

Quantum mechanics and Einstein's special and general theories of relativity formulated at the beginning of the twentieth century are among the most important achievements of human intellect ever. These developments, on the one hand, gave us a clear understanding of what happens inside an atom and a nucleus and, on the other hand, increased our understanding of the evolution of the universe. These ideas are highly counterintuitive and not easily accessible to someone without a sophisticated background in physics and mathematics. Professor Suhail Zubairy's book, *A Mysterious Universe*, brilliantly brings these laws of nature and their consequences to a person with no background in mathematics and physics. This book explains deep and profound concepts in a very simple language. It is rare to find a popular book that covers such a vast amount of material that includes not only the extraordinary laws of quantum mechanics and the amazing laws of relativity but also the astonishing developments in the field of cosmology in such a concise and accessible way. A remarkable feature of *A Mysterious Universe* is that the laws that govern this universe at the micro as well as the macro level are explained with many simple and understandable examples. It is a marvelous book!

David M. Lee, Nobel Prize for Physics 1996

Preface

For thousands of years, since the dawn of civilization, the universe appeared to humans to be quite mysterious. There were so many phenomena that appeared incomprehensible and supernatural. Then the modern scientific revolution started in the sixteenth century. The curtain started lifting from all the mysteries and gradually the laws of nature started to unfold. By the end of the nineteenth century, a feeling started to emerge that all the laws of nature were understood, and these laws could, in principle, explain all the happenings on earth as well as in the cosmos.

Then, at the beginning of the twentieth century, two revolutions brought about the realization that the universe is being run according to the laws that were fundamentally different from the laws that were formulated within the seventeenth, eighteenth, and nineteenth centuries through the work of Galileo, Newton, Young, Maxwell, and others. These revolutions, quantum mechanics and the theory of relativity, that developed around the same time, not only overturned these laws but changed our outlook about the universe. The universe is no longer considered deterministic, particles and waves are no longer exclusive traits of the objects around, the universe had a beginning, reality is no longer objective, and space and time are warped. These are just some of the implications of the new laws of nature.

It is ironical that the universe that was thought to be very well understood and in line with our common sense has become more mysterious than ever. The truth is no longer what we perceive but is much deeper and sometimes incomprehensible.

Both quantum mechanics and relativity are highly mathematical subjects and are not easily accessible. In 2020, I wrote a book *Quantum Mechanics for Beginners* (Oxford University Press, 2020) with the aim of introducing the fundamentals of quantum theory to someone with elementary knowledge of physics and algebra. Here I go one step further and introduce these ideas for someone with no prior knowledge of physics and mathematics.

The purpose of this book is to introduce the fundamental laws of quantum mechanics, relativity, and cosmology to a lay person in as simple a language as possible. The goal is to convey to the reader how strange and mysterious this universe is. Our cherished ideas about space, time, matter, and reality are nothing like what they appear to be. Anyone curious about the foundations of quantum mechanics and the theory of relativity and their consequences would benefit by reading this book.

This book is intended for readers, young and old, who would like to understand the incomprehensible laws that govern the universe without any prior background in physics and mathematics. This book may also be suitable for those with a physics background who would like to glean through the foundational issues of modern physics which are rarely discussed in typical physics courses.

In the first part of the book, topics like wave–particle duality, the probabilistic nature of measurement, the possibility of multiple universes, and the nature of reality are discussed. In the second part, Einstein's special and general theories of relativity and their amazing and mind-boggling consequences are presented. The impact of the theory of relativity on cosmology is immense. The big bang model of the universe, black holes, and the current hot topics of dark matter and dark energy are explained and discussed. These fields that may hold the key to many unanswered questions about the universe are still evolving.

There are a number of people that should be acknowledged for their contributions in completing this book. My deepest gratitude goes to Marlan Scully who has remained an inspiration throughout my professional life. This book is dedicated to him for being a steadfast friend for well over 40 years.

I benefitted from the input of a number of colleagues and friends during the preparation of this book. I am particularly grateful to Bob Brick, Wenchao Ge, Yusef Maleki, Khalid Sohail, and Alexei Sokolov for reading parts of the manuscript and for making helpful suggestions. Special thanks are due to Sonke Adlung and Manhar Ghatora at OUP, Karthiga

Ramu at Integra, and Julian Thomas (copyeditor) for all their help in the publication of this book.

The love and affection received from my family members, Sarah, Neo, Sahar, Shani, Raheel, and Reema, have always been a source of great support. My recent years have been brightened by the loveliest persons in my life: Zoya, Aliya, Qasim, Sameer, Khalid, and Nisa. The person whose support and encouragement has remained the biggest source of strength is my best friend and my wife, Parveen.

M. Suhail Zubairy
College Station, Texas
January 12, 2023

Contents

1

Introduction

*Not only is the universe stranger than we think, it is stranger than we **can** think.*

— *Werner Heisenberg*

Since the dawn of civilization, humans have pondered about the universe around them. They have wondered about the motion of the cosmic objects like the sun, the moon, the planets, and the stars. They have tried to explain the nature and the constituents of objects around them. They have tried to understand how we are able to see and perceive things.

The march toward an understanding of the basic laws of nature has been slow and tedious. As examples, it took almost 2000 years to debunk the Aristotelian idea that all objects are made of four elements, earth, fire, air, and water and realize that they are made of indivisible tiny objects called atoms; it took almost 1000 years before vision was understood as not being through the striking of light rays emitted by the eye on the object to perceive things such as its color, shape, and size, but as the light scattered from the objects into the eye; and it took about 1500 years to move from a geocentric model of the planetary system (in which everything including the sun, the moon, the planets, and the stars revolved around the earth) to a heliocentric model (in which the sun is at the center and the planets, including earth, revolved around it).

Through the millennia, some common observations like thunder and lightning in the sky and solar eclipses when earth becomes dark during the day appeared very mysterious and a lot of myths grew out of their observances. The cover of mystery started lifting with the advent of science and scientific thought. The modern era of science that started in the sixteenth century brought rational thinking to the fore and led to a belief that every phenomenon in this world and the cosmos should have a rational and scientific explanation.

At the end of the nineteenth century, the universe appeared to be completely understood from a scientific point of view. The main conclusions were:

- We live in an infinite three-dimensional universe that has existed for ever.
- Time is completely independent of space and flows uniformly from the past to the present, and then onto the future.
- All the objects in this universe obey laws that are completely deterministic. If we apply a force, any kind of force, we can predict the response very precisely.
- Light is a wave and a ball is a particle—there is no doubt about it. Light cannot behave like a particle and a ball cannot behave like a wave.
- There is no limit to how fast we can move. It is just a matter of building suitable machines and devices that would take us far, very far, at speeds with no limit.
- All the matter in the universe consists of the smallest particles, called atoms. These atoms are like small solid spheres and all the objects are made by stacking these atoms one on another.
- The ultimate vacuum is where there is no movement whatsoever and no energy present.
- All objects that we see around us are real—they continue to exist even when we do not look at them.
- There is only one universe that we see around us. There is no possibility that there are other universes where we may exist in different states (if we are smiling here, we may be sad in another universe).

After the passage of well over one century, our perception about the universe remains the same. All these conclusions remain ingrained within us and any deviation from these would appear to be a complete surprise, indeed shocking, to us.

The amazing fact is that none of these conclusions that were reached through scientific research spread over three centuries ending at the end of the nineteenth century is correct according to the laws of physics that we understand today. All these self-evident truths were shaken and overturned as a result of a scientific revolution that took place in the first quarter of the twentieth century. It was like a massive earthquake whose tremors are felt still today.

The purpose of this book is to convey the fundamental laws of physics as embodied in the quantum and relativity revolutions in as simple a language as possible. This deeper understanding of the laws of nature and the cosmos leads us to ponder that we live in a universe that is incomprehensible. It is truly a mysterious universe.

What was found at the beginning of the twentieth century was that the laws of physics that were formulated by Newton, Young, Maxwell, and others were good only for big objects, intense light, and objects moving very slowly as compared to the speed of light. For small objects like electrons and atoms, very weak light signals, and objects with very high speeds, these laws fail miserably. For example, light can behave both like a wave and a particle. Similarly, an atom can also behave as both a particle and a wave. And someone moving at speeds close to the speed of light lives much longer as compared to a person at rest.

At the dawn of the twentieth century, two revolutions took place independently of each other. The first revolution was entirely due to Albert Einstein, who formulated his theories of relativity during the first twenty years of the twentieth century. These theories would revolutionize our understanding of space and time in ways that were contrary to our common sense. This work had far-reaching consequences in understanding the birth and evolution of our universe.

The second revolution was the birth of quantum mechanics whose formulation took about 30 years starting in December 1900. Quantum mechanics provided the laws that govern the motion of the objects and their interactions with each other and these laws are nothing that we can imagine in our everyday life. They challenge how we perceive our long-cherished concept of reality.

In spite of the highly counterintuitive nature of these theories, quantum mechanics and the theory of relativity are perhaps the two most successful theories in human history. The justification for this remarkable claim is that, after a passage of over 100 years, no physical phenomenon has been found to be in violation of the predictions of these theories. This is true in spite of the tremendous advances in the precision with which the measurements can be made. For example, time can be measured with an accuracy of a billionth of a billionth of a second, distance to a trillionth of a meter, temperature to a millionth of a kelvin,[1] and weight to a billionth of a gram. We can see and manipulate a single atom and cool a gas to an extent that atoms and molecules lose their identity. We can carry out experiments where light consists of a single "photon" and even manipulate the interaction of a single "photon" with a single atom. In all such experiments, the results are dramatically different from what the physics of the nineteenth century predicts but they are remarkably in full agreement with the predictions of quantum mechanics.

In this book, we first give a brief history of how classical mechanics evolved into quantum mechanics. We then attempt to present the foundational issues of quantum theory in a language that should be understandable even to non-physicists. In the latter part, we will discuss Einstein's special and general theories of relativity as well as our understanding of the universe in light of these developments, again in layman's language.

---

[1] Kelvin is a temperature scale similar to the centigrade, shifted by 273 degrees. For example, 0°C is equal to −273 Kelvin, 10°C is equal to −263 Kelvin, and so on. The kelvin scale (K) is chosen such that 0 K is the lowest temperature possible.

PART 1
HISTORICAL

2

Newtonian mechanics and the deterministic world

The seeker after truth is not one who studies the writings of the ancients and, following his natural disposition, puts his trust in them, but rather the one who suspects his faith in them and questions what he gathers from them, the one who submits to argument and demonstration.

— *Ibn al Haytham*

Ulm is the birthplace of Albert Einstein. The most prominent landmark in the city is a church called Ulmer Munster with one of the largest and majestic steeples in the world. During the second world war, the church was badly damaged. However, like other places, Germans reconstructed the church to the extent that a visitor could not sense how badly it was damaged until she saw the pictures of the church taken just after the war. During the reconstruction, they had replaced some painted glass windows that had been destroyed. One painting is particularly interesting. It has the pictures of Copernicus, Kepler, Galileo, Newton, and Einstein (Fig. 2.1). This is highly unusual to see the pictures of scientists in a church where one expected only to see the pictures and statues of Christ, his disciples, and the Christian saints. Even more remarkable is the selection of the scientists in this painting. After all Germany is the country where many other great scientists were born. What was so special about Copernicus, Kepler, Galileo, Newton, and Einstein? Why were such giants of Western science like Faraday, Maxwell, Heisenberg, Schrödinger, and Dirac ignored? A careful consideration reveals that there is one thing common to them: they all explained the laws that dealt with the cosmic objects. Humans have, since antiquity, looked up at the

Fig. 2.1 Glass windows at Ulmer Munster in Ulm.

sky and wondered about the mysterious shining stars and yearned for a better understanding of these elusive objects.

These choices are justified from another angle. In December 2000, it was not just the year that was coming to an end, a century and a millennium were coming to a close as well. *Time* magazine was searching for the Person of the Century, the twentieth century. They wanted to name the person who had the greatest influence on humanity during the preceding hundred years. There was also the question as to who the most influential person of the last thousand years was. There were many distinguished names to choose from all sorts of fields—politicians, poets, writers, philosophers, reformers, conquerors, and scientists. It was remarkable that, in such a tough competition, Einstein was chosen as the Man of the Century and Newton was named the most influential person of the last thousand years.

We are justified in asking why Newton is perceived as an iconic figure of history? What did he do to earn the reputation of the most influential person of the last millennium?

2.1 Newton and the laws of motion

Sir Isaac Newton was born an orphan in 1642, the same year Galileo Galilei died. His transformational work ushered us into the modern era of science and laid the foundations of the industrial revolution of the eighteenth and the nineteenth centuries. He contributed to all branches of mathematics and invented calculus, that provided tools to solve intricate problems in all branches of science. His impact is not diminished by the fact that a German mathematician Gottfried Leibniz also invented calculus independently and around the same time as Newton. Newton made important contributions to optics. However, his greatest contribution lies in formulating universal laws that govern our universe. The hallmark of Newton's discoveries was determinism—we could predict with an arbitrary accuracy the future if we knew all about the past and present.

In order to appreciate the impact of his contributions, one has to go back about a hundred years to the era of Nicolaus Copernicus. In 1543, Copernicus, a monk in a monastery in Poland, presented a heliocentric model of the planetary motion proclaiming the sun at the center and all the planets, including earth, revolving around it in circular orbits. This work challenged the long-held belief that the earth was at the center of the universe and all the planets, the sun, the moon, and the stars revolved around it. This view, going back to antiquity, affirmed the supreme status of human beings at the center of the cosmos. This central status was adopted by the Christian church as well as other religions as a foundational belief that humans were the supreme beings for whom the entire universe was created. Copernicus' model took away this centrality and made earth look like any other planet. More than that, it was inconceivable that humans, and all the other objects, could maintain their stability on a moving earth. They should fall away from the planet with nothing to hold them. The opposition to the heliocentric model was so great that Copernicus could not dare to publish his book for a long time in fear of reprisal from the Church. According to one legend, he received the published copy of his book on the last day of his life, thus dying without knowing the impact that his work would have on subsequent history.

Another breakthrough came through the work of Johannes Kepler. He analyzed the known stellar data and concluded that the orbits of the planets were not circular as proposed by Copernicus, but elliptical. He could empirically derive certain laws of planetary motion based on these observations. For example, he could show that the planets moved slower when farther away from the sun. He could quantitatively derive the speed of the planets in terms of the distance from the sun. This was truly amazing how Kepler could formulate these laws of planetary motion by sifting through the astronomical data.

The next major figure is Galileo Galilei who was born in Pisa in 1564. He is regarded as the father of modern physics. He was the first one to insist that the laws of nature should be written in the language of mathematics instead of a verbal and qualitative account of physical phenomena. He did not invent the telescope but was the first one to use it to observe many stars that were not visible to the naked eye. He discovered the moons of the planet Jupiter. His major astronomical discoveries using telescopes paved the way for the acceptance of the heliocentric system suggested by Copernicus. He made important contributions to the science of motion, discussing the law of falling bodies and parabolic trajectories. Galileo's work preceded the ground-breaking discoveries made by Isaac Newton who was born in the same year (1642) as Galileo died.

In the year 1666, when Newton was a 24-year-old student at Cambridge University, a plague swept the British Isles. The universities were closed for one year and the students were sent home. Newton went to his village where he continued to carry out his research. It was during this period that he made the scientific discovery that heralded the birth of the modern scientific age—he discovered the law of gravitation. The scientific revolution inspired by this discovery continues to this day. Newton discovered that there is always a force of attraction between two massive objects, no matter how small or how big they are. The force is proportional to how massive those objects are. This force becomes small when the masses drift apart.

Newton's law of gravitation could, on one hand, explain why all objects on earth are attracted to the center of the earth, thereby falling on earth, and, on the other hand, explain the planetary motion around the sun and derive the Kepler's empirical laws. This was truly scintillating and absolutely fascinating. This was the first time that a law of nature had been discovered that could explain, in a unified manner, the motion of a small object like a falling apple on earth as well as the details of the motion of planets around the sun. Never before in human history, had a scientific law been stated that could be applicable over such a wide range of objects. Newton's law of gravitation set up the example of how scientific laws should be formulated—they should have universal validity and not restricted to one or few observations.

If Newton had only discovered the law of gravitation and done nothing else, his name would have continued to be among the most influential scientists of human history. But Newton made several other almost equally earth-shattering contributions with lasting impact. He formulated three laws of motion that governed the motion of any object, again as small as a tiny speck and as big as the biggest planet or star in the sky, in the presence of applied forces. He presented his laws of motion in the book *Principia Mathematica Philosophiae Naturalis* in 1686. This book, popularly known as *Principia*, is one of the most influential books ever written (Fig. 2.2).

According to the first law, an object at rest or in uniform motion remains in that state unless an external force is applied on it. Thus, if an object like a cup is placed on a table and no force is applied on it, it will stay at rest in the same position forever. Similarly, if a spaceship acquires a certain speed when its engines are turned off, it will continue to move with the same speed and in the same direction forever unless it comes close to a massive object like a planet which can exert a force of attraction. This was in contrast to the views of Aristotle, according to which an object remains in motion as long as a force is applied on it and as soon as the force is removed, it stops moving. Then why is it that a car moving with a certain speed comes to rest after travelling some distance if we remove our foot from the gas pedal? In this case, there is a force of friction that slows down the car and eventually stops it.

Newton's second law is perhaps the most influential of any law of physics. It states that if an object is at rest or in uniform motion, and a force is applied on it, it experiences an acceleration, that is, the speed changes. The acceleration is directly proportional to the applied force and is in the same direction as the force.

Finally, the third law states that, when one body exerts a force on a second body, the second body simultaneously exerts a force equal in magnitude and opposite in direction on the first body. Thus, if a person tries to push a wall, the wall pushes the person back. To every action there is reaction. A classic example of the application of Newton's third law is

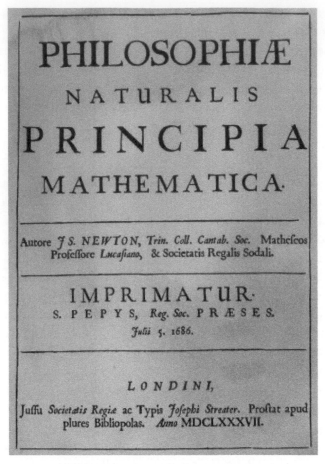

Fig. 2.2 Cover of *Principia* by Isaac Newton.

the motion of a rocket. A rocket engine produces a thrust in the forward direction by exhausting gases that flow in the backward direction.

Newton's laws of motion appear obvious and trivial to us now but these were revolutionary ideas in his time. It was quite new to come up with laws that could be universal and could be applied to everything that existed in the universe. They explained all the known phenomena at that time. This was stunning success. For the first time, some simple laws could not only explain all the existing observations but also predict precisely how the systems will behave under the action of a force, any kind of force. Newtonian mechanics ruled supreme for two centuries until the end of the nineteenth century.

Newton had a great impact outside the science as well. His laws of motion took away mystery from many phenomena that had remained elusive for thousands of years. The era of science that was unleashed by Newton's discoveries led to a realization that all natural phenomena must have a scientific explanation.

2.2 Failures of Newtonian mechanics

There were, however, implicit in Newtonian laws certain predictions that did not pass the test of time. This led to a replacement of these laws with more fundamental laws at the beginning of the twentieth century, relegating the Newtonian mechanics to being a good approximation for objects that are sufficiently large and moving at small speeds.

An important consequence of Newton's laws of motion is that the motion of a particle is deterministic: If we know the initial position and speed of the particle as well as all the forces acting on it, then we can predict with certainty its location and speed at a subsequent time with arbitrary precision. In other words, the trajectory of the particle can be traced in advance. Another way to look at it is that, if we know the precise location of each object in this universe and we also know all the forces acting upon them, then the future evolution of the universe can be precisely determined.

As we discuss in a later chapter, Newtonian laws failed miserably when applied to small objects like atoms. A search of new and resilient set of laws led to the birth of quantum mechanics at the dawn of the twentieth century. Quantum mechanics has replaced the Newtonian determinism with the probabilistic nature of our observations.

A hallmark of the Newtonian mechanics is the absolute nature of space and time.

Newton described the notion of time in these words: "... absolute and mathematical time, of itself, and from its own nature, flows equally without relation to anything external." Space and time are therefore distinct. Two events happening at two different points in space will be perceived as simultaneous if they take place at the same absolute time. Newton's concept of an absolute and mathematical time applies for all observers at all places equivalently—for example, three friends, one at home, the other in a train and another traveling in an airplane, can have lunch at the same "absolute" time. Time flows continuously and regularly from past to present and onto the future. The flow of time cannot be affected by anything—gravity, heat, force.

"Absolute space, of its own nature without reference to anything external, always remains homogeneous and immovable." This is how Newton described the absolute nature of space—three-dimensional space. The space is, however, different: contrary to time which flows in one direction, it is possible to move in space at will. We can move in one direction and, after travelling some distance, can retrace our steps and return to the starting point. The length of an object is the same no matter whether the object is at rest or moving with a high speed.

Another consequence of Newtonian laws is that there is no limit on how fast an object can move. In principle, if a constant force is applied to an object for a very long time, the speed of the object can be increased to an unlimited value.

These conclusions, regarding the absolute nature of space and time as well as the possibility of an unlimited speed for an object, were challenged by Albert Einstein through his theory of relativity, with mind-boggling consequences. Einstein's work changed the way we look at space and time forever. Our perception of flat space with absolute time that was so successfully formulated in Newtonian mechanics and is ingrained in us was shattered by Einstein's work. The nature of the universe around us is not what it appears to be. At a fundamental level, the laws that govern the universe are truly strange and bizarre, as we learn in later chapters.

3

Quantum mechanics is born

Classical physics could explain the world but got some of the details wrong; quantum physics gets all the details right but can't explain the world.

— *Bruce Rosenblum*

At the end of the nineteenth century, the phenomena of mechanical motion, electricity and magnetism, thermodynamics, and, light could be explained in a satisfactory manner by the scientific laws as discovered by Isaac Newton, Thomas Young, Rudolf Clausius, Michael Faraday, James Clerk Maxwell, and others, and there was justification in feeling that the basic laws of nature were fully understood. So much was the satisfaction with the existing laws of physics that a very eminent British scientist, Lord Kelvin, is quoted as saying in an address to the British Association for the Advancement of Science in 1900, "There is nothing new to be discovered in physics now. All that remains is more and more precise measurement." There were, however, a small number of unresolved problems at the dawn of the twentieth century that could not be explained on the basis of the existing theories. A resolution of these problems led to a major revision of the existing laws that in turn led to the birth of two revolutionary theories: quantum mechanics and the theory of relativity. The objective of this chapter is to review the steps that led to the development of quantum mechanics. The discovery of the theory of relativity will be discussed in a later chapter.

The development of quantum mechanics, that replaced the classical mechanics of Newton and Maxwell, took place in two distinct eras. First, we discuss the era between 1900 and 1925 when certain phenomena that could not be explained by the known laws of physics required a quantum hypothesis for their explanation. This was, therefore, a period of unprecedented crisis in the history of physics, when the foundations of physical theories built by Newton, Young, Maxwell, and others over the centuries were crumbling and no new theory was there to replace them. Then the clouds started clearing in the summer of 1925 when the laws of full quantum theory, laws that would replace Newton's laws, started taking shape. In the second part we discuss the salient features of these developments.

3.1 Max Planck and blackbody radiation

The first era began in December, 1900, when Max Planck introduced the notion of the quantization of energy to explain the color distribution of light radiated by hot objects. This mundane problem had remained unresolved for almost 40 years. A puzzle confronting physicists in 1900 was just how do heated objects radiate? A solid consists of atoms and molecules, and heat causes them to vibrate. However, atoms and molecules are themselves complicated patterns of electrical charges. Oscillating charges emit light. The picture, then, is that when an object is heated, the consequent vibrations on the atomic and molecular scales inevitably induce charge oscillations. These oscillating charges radiate, giving off heat and light that are observed.

In 1859, a German physicist, Gustav Kirchhoff, addressed this problem. Central to Kirchhoff's studies was the concept of a *blackbody*, an object that absorbs all the radiation that falls on it. In practice there is no object that is ideally black, but many objects in the real world come close to exhibiting blackbody behavior. A perfect blackbody can also emit radiation with a certain color distribution at a given temperature. Kirchhoff proved by general thermodynamic arguments that the color distribution of the emitted radiation from a blackbody depends only on its temperature and is independent of the material. Kirchhoff posed it as a challenge to find, for each temperature, the precise color distribution of the emitted light. A search of this distribution would lead to the birth of quantum mechanics, literally at the end of the nineteenth century, on December 14, 1900, by the German physicist, Max Planck.

In order to understand the radiation emitted by the heated objects, we notice that, at room temperature, a metallic object like an iron rod emits radiation that peaks at near infrared, which we call heat radiation. As we increase the temperature, the color of the heated object changes. Each color corresponds to a different wavelength, which is the distance between two neighboring crests. The wavelength is large for red and decreases as the color gradually changes to orange, yellow, green, and blue, as shown in Fig. 3.1. Another related quantity is the frequency which is the number of crests passing through a point in one second. Since the speed of light is the same for all the colors, the frequency is lowest for the color with the largest wavelength. Thus, the frequency is low for red light and increases as we move toward blue as shown in Fig. 3.1.

The shifting of the emitted radiation across the rainbow colors as the temperature increases is a well-observed phenomenon. For example, when an iron rod is heated in a furnace, the color of the iron rod first turns dull red and on further increase in temperature it changes to bright red to orange, then yellow, progressively white, and then finally blue at the elevated temperatures. Experimentally, it was observed that the heated objects emitted a whole distribution of colors around these peaks, as shown in Fig. 3.2. It was a challenge to explain this color distribution using the known laws of physics. For almost forty years this problem

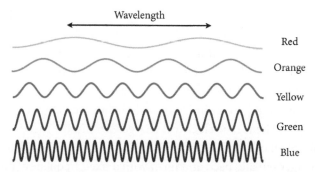

Fig. 3.1 Waves with increasing frequency or decreasing wavelength for different colors from top to bottom. Red color has the largest wavelength (distance between two adjacent crests) and blue has the shortest wavelength.

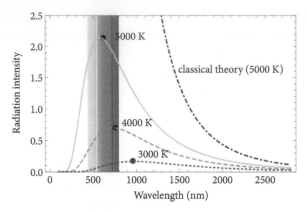

Fig. 3.2 The spectrum of light emitted by a hot object. The peaks shift from smaller wavelength to higher wavelength as the temperature decreases. The classical theory predicted an infinite radiation at low wavelengths or higher frequencies such as for ultraviolet radiation. This is referred as "ultraviolet catastrophe."

remained unresolved. The theories of physics at that time were unable to solve this problem.

By 1900, this failure had caused people to question the correctness of the known theories of classical physics and thermodynamics. It was, however, Max Planck (1858–1947) who eventually presented the radiation formula that matched the experimentally observed color distribution at a given temperature. Planck presented his results that would eventually revolutionize our understanding of the laws of nature in ways that even Planck could not have anticipated at that time.

When Planck addressed the problem of heating the metallic object, he realized that, since the results were independent of the nature of the material, one could use a simple model for the oscillation of atoms and molecules, and hence the electrical charges, within the metals. He chose a very simple model for these oscillations: a harmonic oscillator which, like a swing, oscillates back and forth with a fixed frequency (number of oscillations per second).

The behavior of a swing is fully understood. If we give a strong push, a swing moves back and forth coming to rest at the extreme points and

Fig. 3.3 A swing can move back and forth and executes a periodic motion with no restriction on the amplitude of oscillation. Quantum mechanics allows only certain amplitudes and not others.

(Credit: Sahar Zubairy)

moving fastest in the middle (Fig. 3.3). The motion is completely periodic meaning that each complete cycle takes the same amount of time, regardless of how far the swing moves. The number of cycles per second is called the frequency. The frequency of the swing can be changed by changing the length of the rope and nothing else. An important point is that there is no restriction, absolutely no restriction, how far the swing moves. It can be a small distance or it can be a large distance from midpoint. The energy of the swing depends on how far it can move—more energy is stored in the swing if it moves to a farther point. Since there is no restriction on how far the swing can move, there is no restriction on the amount of energy that can be stored in the swing.

In Planck's simplified model, the electric charges in the atoms and molecules of the material oscillate just like a swing, and consequently emit light. However, when Planck assumed this reasonable behavior for the atoms and molecules, he got results that did not agree with the experimental observations.

Then, Planck did something that was apparently completely unreasonable and a departure from what physicists had known until that time.

This act, that Planck himself described as "an act of desperation," would change the course of the science of the twentieth century.

He assumed that the oscillators cannot have any arbitrary energy. Instead, the energy comes in packets or quanta. This is like saying that the swing can move to only certain points and not others. Thus, for a given frequency, the smallest energy packet will have an energy equal to a constant multiplied by the frequency of the oscillator. Mathematically we can write it as $E = hf$ where E is the energy of the packet, h is a universal constant which is named Planck's constant in honor of Max Planck, and f is the frequency (number of oscillations per second) of the oscillator. The only other allowed energies are 2 times E, 3 times E, and so on as shown in Fig. 3.4. According to Planck's hypothesis, the oscillator cannot have 3/2 times E or 2/3 times E etc. as allowed energies. They are forbidden. With this assumption, Planck could derive the emitted color distribution that matched the experimental results. This was a stunning success but at the price of abandoning the common-sense behavior of the oscillators.

The value of Planck's constant h is extremely small, equal to $6.62607015 \times 10^{-34}$ meter^2 kilogram/second.[1] As we shall see in Section 3.7, the smallness of Planck's constant is the reason why we do not see quantum effects explicitly for big objects in our everyday life. The quantum effects are more important at the level of atoms and electrons.

At the time he proposed this radical hypothesis, Planck could not explain why energies should be quantized. However, his hypothesis solved the long-standing problem of explaining the blackbody radiation spectrum with amazing success.

Fig. 3.4 The energy of a harmonic oscillator is quantized in units of hf where h is Planck's constant and f is the frequency of oscillation.

---

[1] $10^{-34} = 0.0000000000000000000000000000000001$

Planck's hypothesis to explain the light emitted by heated objects was proposed in 1900. It was a revolutionary idea that energy should come in packets or "quantized." However, it was not perceived as such at the time it was proposed. When Planck proposed his theory, there was no dancing on the streets or major headlines in the newspapers. Even the scientific community at large did not grasp the significance of the quantization condition.

For almost five years, Planck's hypothesis could not find any application until Albert Einstein used the quantum condition to explain another unexplained effect, the photoelectric effect, and introduced the notion of a "photon" in his well-known Nobel Prize winning paper of 1905.

3.2 Einstein and the photoelectric effect

In the 1890s, Heinrich Hertz (and later Philipp Lenard) observed that when a light beam is incident on a metallic surface, electrons are ejected from the surface (Fig. 3.5). The model that was used to understand this phenomenon was that electrons are part of the atom (we present the model of the atom known at the turn of the twentieth century in Chapter 5) and if they are provided with sufficient energy, which is different for different metals, these electrons, called photoelectrons, will be released and leave the metal with some energy.

It was however observed that, for certain colors like red, no photoelectrons were emitted, no matter how intense the light beam (Fig. 3.6). However, for other colors, like blue and violet, the photoelectrons were emitted no matter how weak the light beam. For such light beams, the emission of photoelectrons takes place almost instantaneously after the

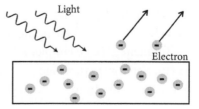

Fig. 3.5 Photoelectric effect: When light is incident on a metal, photoelectrons are emitted.

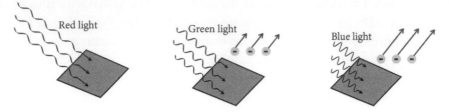

Fig. 3.6 No electrons are emitted if the frequency of light is below a critical frequency. However, as the frequency of light is increased further, electrons are emitted with increasing speed.

light shines on the metal, with no detectable time delay even if a very low intensity of light is incident.

These observations were extremely surprising and could not be explained on the basis of classical laws of physics known at the end of the nineteenth century. For example, how could it be that photoelectrons are not emitted for certain colors of light even if the light intensity is very large but then photoelectrons are emitted even with a feeble light beam for some other colors? And then the most mysterious effect was the instantaneous emission of photoelectrons even when a very weak beam of light was incident on the metal.

The picture of light at that time was that it consisted of waves and if sufficiently intense waves of light were incident on the metal, electrons should be ejected. It may take some time before these waves accumulate enough energy but when this energy is reached, it can be imparted to the electrons leading to the photoelectron emission. However, this wave picture was unable to explain the observed behavior—there was no reason why waves of certain colors should be able to eject an electron and other colors could not.

In 1905, Einstein explained the photoelectric effect using Planck's hypothesis of quanta of energy. Einstein assumed that light behaves like a collection of particles or quanta, called photons, with each photon carrying an energy depending on the color. Each color is identified by its frequency. For example, a red photon with smaller frequency has smaller energy than a blue photon. The photon energy increases in the rainbow colors from red, orange, yellow, green, blue, indigo, to violet.

When one of these photons penetrates the metal, it gives all its energy to the electron.

Einstein assumed that each photon "collides" with an electron and imparts its energy to the electron. If the energy of the photon is greater than a certain minimum value for a given metal, the electron is ejected. This explains why, for a certain metal, a beam of red light could not emit the electron no matter how intense the light is. None of the red photons, even if they are very large in number, has enough energy to eject the electron. On the other hand, a blue color photon, even if there is only one, has sufficient energy to force the electron out. This explains the instantaneous emission of photoelectrons.

An analogy can explain this behavior. Suppose, we wish to get an unwilling person to be ejected from a room. If a large army of Lilliputians is sent, they would not succeed.[2] But a single powerful person, like Gulliver, should be sufficient to eject the person from the room. In Einstein's explanation, low frequency red photons play the role of Lilliputians and the high frequency blue photon plays the role of Gulliver (Fig. 3.7).

So, in a single stroke, Einstein explained the photoelectric effect. He postulated a "particle" type picture for light and introduced the notion of "light quanta" or photons. This was a bold move.

Einstein's explanation of the photoelectric effect was the first vindication of Planck's hypothesis. It was also the first time that light quanta were introduced. The idea that light consists of photons had a great impact on subsequent developments in the full formulation of quantum theory. The concept of photons explained the photoelectric effect so beautifully. However, the picture of light as consisting of photons could not explain some other well-known phenomena such as interference and diffraction. As we discuss in Chapter 4, only a wave nature of light could explain these phenomena.

Now there was a dilemma. On the one hand, phenomena like the photoelectric effect could only be explained if light is treated like a particle (photon) and could not be understood if it is treated like a

---

[2] Lilliput is a fictional island in the 1726 novel *Gulliver's Travels* by Jonathan Swift. The island is inhabited by tiny people who are about 4–6 inches tall.

Fig. 3.7 The Emperor of Lilliput and his army marching between Gulliver's legs.

(From *Gulliver's Travels* by Jonathan Swift, originally published in 1726.)

wave. On the other hand, phenomena such as interference could only be understood by treating light as a wave, but could not be explained in the "photon" picture. Thus, we have a paradoxical picture of light: in some experiments it behaves like a wave and, in others, it behaves like a particle. This was the first time that the mysterious wave–particle duality behavior entered in our understanding of the laws of nature.

This was a dilemma whose complete resolution, via a formal theory that would rigorously explain all these phenomena within the structure of a single theory, had to wait almost a quarter century—till the birth of quantum mechanics in the summer of 1925. How light knows when to behave as a wave as in an interference experiment and when to behave as a particle as in the photoelectric effect has, however, remained a perplexing question for over a hundred years. We discuss some counterintuitive aspects of the wave–particle duality in Chapter 10.

3.3 Niels Bohr and the hydrogen atom

The third problem relates to the atomic structure of hydrogen. It was known through the work of Rutherford that an atom consists of a positively charged nucleus with negatively charged electrons orbiting around it. It was also known that hydrogen atoms produced some discrete spectral lines corresponding to light energy radiated at some well-specified frequencies. How to reconcile Rutherford's model with the observation of the spectral lines? This was a challenge around 1913 when Bohr applied a quantization condition to explain this curious effect. Bohr's condition implied that the electron can exist only in some well-defined orbits whose radii can be obtained from the quantization condition. The spectral distribution of the radiation emitted can then be explained by assuming that, when an atom in a higher orbit jumps to a lower orbit, a photon of discrete frequency is emitted. Bohr could show that the frequencies of the emitted light predicted based on the quantization condition matched the experimentally observed frequencies. Some more details of the Rutherford and Bohr model will be presented in Chapter 5.

Thus, the theme of quantization of energy, so alien to classical mechanics, became fundamental to the quantum mechanical description. For example, a particle such as an electron moving freely inside a box can lead us to the surprising result that the particle can only have discrete amounts of energy. This is in sharp contrast to our everyday observance that, in the macroscopic world, particles can have any amount of energy.

3.4 Particles behave like waves—de Broglie

In 1905, Einstein explained the photoelectric effect by showing that, in some experiments, light waves act like particles. What about the particles like a tennis ball, a dust particle, or an electron behaving like waves? In 1924, Louis de Broglie argued in his Ph.D. thesis that if light can behave both like waves, as in interference and diffraction, and like particles, as in the photoelectric effect, then particles should also behave like both particles and waves. The de Broglie hypothesis completed the wave–particle duality description of both waves and particles. A wave is characterized by frequency (number of crests per second) or wavelength (the distance between the two crests) and de Broglie argued that a particle moving with a certain speed is characterized by a wave of a certain wavelength, called the de Broglie wavelength. The smaller the speed, the larger is the wavelength.

When de Broglie postulated that particles can behave like waves, there was no evidence to support this conjecture. However soon it was shown experimentally that de Broglie's hypothesis was indeed correct. De Broglie was awarded the 1929 Physics Nobel Prize. In his Nobel acceptance speech, de Broglie described his discovery of de Broglie waves in these words:

On the one hand the quantum theory of light cannot be considered satisfactory since it defines the energy of a light particle (photon) by the equation $E = hf$ containing the frequency f. Now a purely particle theory contains nothing that enables us to define a frequency; for this reason alone, therefore, we are compelled, in the case of light, to introduce the idea of a particle and that of frequency simultaneously. On the other hand, determination of the stable motion of electrons in the atom introduces integers, and up to this point the only phenomena involving integers in physics were those of interference and of normal modes of vibration. This fact suggested to me the idea that electrons too could not be considered simply as particles, but that frequency (wave properties) must be assigned to them also.

The de Broglie hypothesis that particles behave like waves seems quite mysterious. We do not seem to see the particles around us as waves. The particles, no matter how small, are well-defined objects and cannot, for a moment, be perceived as a wave. A baseball, or even a dust particle, cannot be described as waves. Why? The reason is that the corresponding de Broglie wavelength is small, unimaginably small, and therefore the wave nature is completely masked. Only for particles of the size of atom and electrons is the wave nature exhibited.

A big success of de Broglie's hypothesis was, as de Broglie pointed out in his Nobel lecture, that it provides an insight into the Bohr quantization condition and tends to resolve somewhat why certain orbits are allowed and not others. De Broglie showed that, if an electron in the hydrogen atom is described by a wave of a certain wavelength, then only those orbits are allowed for which the circumference of the allowed orbit of an electron is an integral multiple of the de Broglie wavelength of the electron. This is shown in Fig. 3.8. The de Broglie hypothesis perfectly matched with Bohr's quantization condition.

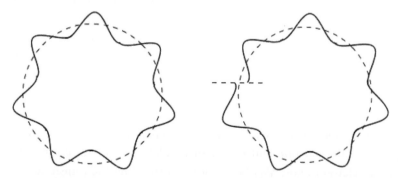

Fig. 3.8 (a) The Bohr quantization condition yields a stable orbit such that the circumference of the allowed orbit of an electron is an integral multiple of the de Broglie wavelength of the electron. (b) When the Bohr condition is not satisfied, a stable orbit cannot be obtained.

Louis de Broglie's hypothesis played an important role in laying the foundation of quantum mechanics.

3.5 Birth of quantum mechanics

These successful attempts to explain some unresolved phenomena based on Planck's quantization hypothesis led to the realization that the old classical theory, as formulated by Newton, Young, Maxwell, and others, may not be valid when we try to understand phenomena at the atomic level. Planck, Einstein, and Bohr could explain some unresolved phenomena based on postulates that involved quantization of energy that had no basis in classical theories. Despite these successes, however, there was no theory that could explain these and all other phenomena in a unified manner. It was becoming apparent with the difficulties being faced in attempting to explain new emerging results at the microscopic level that a full-fledged theory was needed that should replace Newtonian mechanics.

The second era began with the breakthrough that came in the summer of 1925 when 24-year-old Werner Heisenberg took the first major step in formulating a quantum theory, making a clean break with the past. His fundamental idea was to develop a theory that should include, in principle, only those quantities that are observable. The opening words of his epoch-making paper were:

> The present paper seeks to establish a basis for theoretical quantum mechanics founded exclusively upon relationships between quantities which in principle are observable.

For example, energy emitted by an atom is an observable quantity whereas the position and period of oscillation of the electron around the nucleus are not. Max Born soon realized that the transition quantities obeyed the rules of matrix algebra, a branch of mathematics that was not well known at that time. A more complete version of the matrix mechanics version of quantum mechanics was formulated in a paper by Born, Jordan, and Heisenberg.

In January 1926, Erwin Schrödinger independently formulated the quantum theory and wrote down a dynamical equation that is called the Schrödinger equation in his honor. Later it was shown that the theories of Heisenberg and Schrödinger were two different but completely equivalent formulations of quantum mechanics. Schrödinger's equation is one of the most famous equations in physics. We discuss it further in Chapter 7.

Quantum mechanics, as formulated by Heisenberg and Schrödinger (along with other founding fathers including Max Born, Pascual Jordan, Paul Dirac, and Wolfgang Pauli), could not only explain all the existing phenomena at the microscopic and macroscopic levels but also predict new phenomena that could then be observed experimentally. Despite these stunning successes of the new theory, the conceptual foundations of the theory became a major point of discussion. What we see is that, at the level of a single atom or an electron or a photon, quantum mechanics makes predictions that are startling. They are dramatically different from the corresponding results for our everyday objects that can be described very successfully using Newtonian mechanics. The mind-boggling aspect of quantum mechanics was not lost on the founding fathers. Indeed, in spite of the great successes in explaining and predicting novel phenomena, the conceptual foundation of quantum mechanics remains a hotly debated issue. The quantum mechanical laws, as opposed to Newtonian laws, are contrary to common sense and have led to debates on the meaning of physical reality.

3.6 Why we do not see quantum effects in everyday life?

The laws of quantum mechanics replaced the laws of Newtonian mechanics. However, in our everyday life, the mechanical laws of Newton provide amazingly accurate results. There is hardly any observation where Newton's laws appear to fail and we need to resort to quantum mechanics to explain a certain phenomenon. With few exceptions, it is, as mentioned above, at the atomic level that quantum mechanical predictions are at variance with the results obtained using classical mechanics. But why?

The main reason that Newtonian mechanics is extremely good approximation is due to the incredibly small value of the Planck's constant, h, being equal to 0.000000000000000000000000000000000626 kilogram-meter-meter per second. In a compact notation, Planck's constant is written as $h = 6.26 \times 10^{-34}$ kilogram-meter-meter per second.[3]

For a photon of green light, the frequency f is 5.40×10^{14} cycles per second. The energy of a single green photon is equal to the product of Planck's constant and the frequency, which is equal to an incredibly small number, 3.4×10^{-19} joules.

An example of when the quantum mechanical effects can become observable is when an object of a certain mass moving in a circle with a speed is so small that the multiplication of its mass, speed, and radius becomes comparable to the value of Planck's constant. A careful analysis indicates that this places rather stringent requirements on the mass, speed, and the radius of the circular orbit. This can be illustrated with the following examples.

Let us first consider a tired gnat whose mass is 10 milligram moving slowly at a speed of 10 centimeters per second in a circle of radius of 1 centimeter. If we multiple the mass, the speed, and the radius, the product is equal to 0.00000001 kilogram-meter-meter per second. This number, compactly written as 10^{-8}, is a hundred trillion trillion times bigger than Planck's constant 6.62×10^{-34}.

As another example, let us consider a much smaller object, a dust particle, of mass equal to 10^{-12} kilogram moving in a radius of 1 millimeter with a speed of 10^{-4} centimeters per second. The product of mass, radius, and speed turns out to be 10^{-21} kilogram-meter-meter per second. This is an incredibly small number but still more than a billion times larger than Planck's constant.

Finally, we consider an electron of mass equal to 10^{-30} kilograms orbiting around a nucleus in a radius 10 Angstrom (equal to 10^{-9} meters) at a speed of 6.62×10^5 meters per second. Then the product of mass,

---

[3] Here we explain the notations such as 10^{12} and 10^{-12}. The notation 10^{12} describes a number with 1 followed by 12 zeros, that is $10^{12} = 1,000,000,000,000$, which is equal to one trillion. This is a large number. Similarly, 10^{-12} is one divided by one trillion, that is $10^{-12} = \frac{1}{10^{12}} = \frac{1}{1,000,000,000,000} = 0.000000000001$, which is equal to one trillionth. This is a small number.

speed, and radius becomes equal to 6.62×10^{-34} kilogram-meter-meter per second, the same value as that of Planck's constant.

This indicates why quantum effects may not be observable for a small, tired gnat or even a dust particle. However, these effects may become very important at the levels of atoms and electrons. Of course, with an improvement in the measurement tools, the quantum effects for bigger objects may become observable.

4
What is light?

All the fifty years of conscious brooding have brought me no closer to the answer to the question: What are light quanta? Of course, today every rascal thinks he knows the answer, but he is deluding himself.

— *Albert Einstein*

The nature of light has been a subject of interest going back to antiquity. Until around the seventeenth century, studies of light were mainly concerned with vision. For example, the ancient Egyptians believed that light was the activity of their god Ra seeing. When Ra's eye (the Sun) was open, it was day. When it was closed, night fell.

The earliest studies on the nature of light and vision can be attributed to the Greek and Hellenistic traditions. The Greek period, extending from the Archaic period till around 320 BC and centered in Athens, produced many of the earliest ideas about vision through the works of Democritus, Epicurus, Plato, and Aristotle. After the death of Alexander, the center shifted to Alexandria where Ptolemy I, a general in Alexander's army, established a new dynasty that lasted till the Roman conquest of Egypt in the first century BC. In this Hellenistic period, the glorious traditions of Greek scholarship in the field of light and vision continued through the works of Euclid, Hero of Alexandria, Ptolemy, and Galen. Central to the theory of vision was the idea that, just as other senses like touch and taste, vision should be via rays of light emanating from the eye that sense the objects. This theory, called the "extramission" theory, remained influential till Alhazen in the eleventh century who showed clearly that the light rays scattered from objects to the eye are responsible for vision.

A serious study on the nature of light started in the seventeenth century when Isaac Newton proposed that the light consists of small particles. His contemporary Christian Huygens described light as being like waves in a pond. The interference experiment by Thomas Young and the diffraction experiment by Augustin Fresnel in the early nineteenth century debunked the particle nature of light and established light as consisting of waves. Another milestone came in the later part of the nineteenth century when James Clerk Maxwell, while trying to unify two of the fundamental forces of nature, electric and magnetic forces, established light as an electromagnetic wave.

Our concept of light underwent a dramatic change with the advent of quantum mechanics. The quantum mechanical picture of light is the most startling—it behaves like a wave in experiments like interference and diffraction and in others like a particle called a photon. This incomprehensible picture has stayed with us till now.

We have therefore come a long way from the earliest studies on light, trying to understand vision as light emanating from our eyes, to the description of light as rays, then as particles, and then waves, and finally exhibiting both particle and wave natures. We can only speculate how our present understanding of light will be perceived decades or centuries from now. Will our picture of light quanta as both waves and particles survive or will something more intuitive replace this incomprehensible picture? It is an irony that the greatest strides taken in the scientific understanding have come in our time, yet we feel least certain of our understanding of what light is, what a photon is. In spite of the great success of the mathematical theory to describe light and its amazing agreement with experiments, the question "What is light?" can ignite a heated discussion.

4.1 Greeks and antiquity

The theory of vision attempts to explain how objects, near and far, their shape, size and color, are perceived by us. The earliest systematic studies of vision are attributed to atomists who reduced every sensation, including vision, to the impact of atoms from the observed object on the organ of observation. There were different schools of thought among atomists. For example, Democritus (460–370 BC) believed that the visual image did not arise directly in the eye, but the air between the object and the eye is contracted and stamped by the object seen by the observing eye. The pressed air contains the details of the object and this information is transferred to the eye. Epicurus (341–270 BC), on the other hand, proposed that atoms flow continuously from the body of the object into the eye. However, the body does not shrink because other particles replace and fill in the empty space.

An alternate theory of vision due to Plato (428–328 BC) and his followers advocated that light consisted of rays emitted by the eyes (Fig. 4.1). The striking of the rays on the object allows the viewer to perceive things such as the color, shape, and size of the object. Our vision was initiated

Fig. 4.1 Extramission theory of light. Four persons see a dragon during flight by the rays emitted by their eyes.

(Johann Zahn, "Oculus Artificialis Teledioptricus Sive Telescopium," 1685.)

by our eyes reaching out to "touch" or feel something at a distance. This is the essence of the extramission theory of light that would be influential for almost 1000 years until Alhazen would conclusively prove it to be wrong.

Euclid (b. 300 BC) is the father of Geometry. His book *Elements* laid down the foundation of the axiomatic approach to geometry and is one of the most influential books ever written. His work in optics follows the same methodology as *Elements* and gives a geometrical treatment of the subject. Euclid believed in the extramission theory and his theory of vision is founded in the following postulates:

1. Rectilinear rays proceeding from the eye diverge indefinitely;
2. The figure contained by the set of visual rays is a cone of which the vertex is at the eye and the base at the surface of the object seen;
3. Those things are seen upon which visual rays fall and those things are not seen upon which visual rays do not fall;
4. Things seen under a larger angle appear larger, those under a smaller angle appear smaller, and those under equal angles appear equal;
5. Things seen by higher visual rays appear higher, and things seen by lower visual rays appear lower;
6. Similarly, things seen by rays further to the right appear further to the right, and things seen by the rays further to the left appear further to the left;
7. Things seen under more angles are seen more clearly.

Euclid did not define the physical nature of these visual rays. However, using the principles of geometry, he discussed the effects of perspective and the rounding of things seen at a distance. Euclid had restricted his analysis to vision.

Hero of Alexandria (10–70), who also believed in the extramission theory of Euclid, extended the principles of geometrical optics to consider the problems of *catoptrics*,[1] particularly, reflection from smooth surfaces. Hero derived the law of reflection by invoking the principle of

---

[1] *Catoptrics* is the branch of optics that deals with reflection.

least distance. According to him, light from point A to another point B follows a path that is shortest. On this basis, he showed that when light reflects from a surface, the angle of incidence is equal to the angle of reflection. Specifically, the image appears to be as far behind the mirror as the object is in front of the mirror. Hero's principle of least distance would be replaced by the principle of least time by Pierre Fermat more than 1500 years later to derive the law of refraction.

The most influential and perhaps the last important figure in optics of the Greek–Egyptian era was Claudius Ptolemy (90–168). He is most well-known for championing the geocentric model for the movement of planets, a view that would survive for almost 1400 years until it was replaced by a heliocentric model through the work of Nicholas Copernicus in 1543. His book on the subject, *Almagest*, was very influential in shaping the thinking on astronomy and, along with *Elements* by Euclid, is the oldest surviving book in the history of science.

Ptolemy wrote *Optics* in which he discussed the theory of vision, reflection, refraction, and optical illusions. Like Euclid and Hero, Ptolemy championed the extramission theory of vision. He considered visual rays as propagating from the eye to the object seen. However, instead of considering visual rays as discrete lines as postulated by Euclid, he considered them forming a continuous cone. Ptolemy carried out careful experiments on refraction and concluded that, for light propagating from one medium to another, the ratio of the angle of incidence to the angle of refraction was constant and depended on the properties of the two media. He thus derived the small angle approximation of the law of refraction. The formulation of the theory based on experimental results, frequently supported by the construction of special apparatus, is the most striking feature of Ptolemy's *Optics*.

4.2 Alhazen: End of extramission theory

The extramission theory remained influential for almost a thousand years until Alhazen conclusively proved it to be wrong in the beginning of the eleventh century. Alhazen, an Arab scientist, proved, that, contrary to the conventional theory of vision, light originated, not from the eye, but from the illuminated objects.

Abu Ali al-Hasan ibn al-Hasan ibn al-Haytham, known in the west as Alhazen, is a central figure in science. He is often described as the greatest physicist between Archimedes and Newton. He was the first person to follow the scientific method, the systematic observation of physical phenomena and their relation to theory, thus earning the title First Scientist from many.

His most important contribution in optics is his book *Kitab-al Manzir* (Book of Optics) which was completed around 1027. This book, comprising seven volumes, was the first comprehensive treatment of optics and covered subjects such as the nature of light, the physiological treatment of the eye, and the light bending and focusing properties of lenses and mirrors. This book was most influential in the transition from the Greek ideas about light and vision to modern-day optics. Alhazen's Book of Optics was translated in Latin at the end of the twelfth century under the title *De Aspectibus* and would remain the most influential book in optics till Newton's *Opticks* published in 1704.

Alhazen proved the long-held theory of Euclid, Hero, and Ptolemy that light originated from the eye to be wrong and showed that light originated from the light sources. He did this by carrying out a simple experiment in a dark room where light was sent through a hole by two lanterns held at different heights outside the room. He could then see two spots on the wall corresponding to the light rays that originated from each lantern passing through the hole onto the wall. When he covered one lantern, the bright spot corresponding to that lantern disappeared. He thus concluded that light does not emanate from the human eye but is emitted by objects such as lanterns and travels from these objects in straight lines.

Based on these experiments, he invented the first pinhole camera (that Kepler would use and call a camera obscura in the seventeenth century) and explained why the image in a pinhole camera was upside down.

Alhazen's theory of vision was not limited to the description of light rays originating from the objects and entering the eye. He also understood that an explanation of vision must also take into account anatomical and psychological factors. He proved that the perception of an image occurs not in the eyes but in the brain and that the location of an image is largely determined by psychological factors.

Alhazen did not invent the telescope but he explained how a lens worked as a magnifier. He contended that magnification was due to the bending, or refraction, of light rays at the glass-to-air boundary and not, as was thought, to something in the glass. He correctly deduced that the curvature of the glass, or lens, produced the magnification. He concluded that the magnification takes place at the surface of the lens, and not within it. His work on catoptrics in Book V of the Book of Optics dealt with problems of reflection from spherical and parabolic mirrors.

4.3   From Johannes Kepler to Isaac Newton

The most important figure to follow Copernicus was the German astronomer, Johannes Kepler, whose laws on planetary motion would prove pivotal to the discovery of Newton's law of gravitation. Kepler is a key figure in the history of light and vision as well. His interest in the subject appears to have originated in his observation of a solar eclipse on July 10, 1600 by means of a camera obscura. Several years before that, Tycho Brahe, the greatest naked-eye astronomer of the time, had observed that the angular diameter of the moon appeared to be larger during a solar eclipse when observed through the pinhole camera than when observed directly. Kepler understood that this anomaly could not be explained without a full understanding of the optical instruments, in this case, the camera obscura. He noted that the finite diameter of the pinhole should be responsible for this anomaly. He discovered the solution by an experimental technique where he stretched a thread through an aperture from a simulated luminous source to the surface on which the image was formed. He traced out the image cast by each point on the luminous body seeing, in the process, the geometry of radiation in three-dimensional terms. In this way, Kepler was able to formulate a satisfactory theory of radiation through apertures based on the rectilinear propagation of light rays. Kepler did not stop at explaining Tycho Brahe's problem of seemingly variable lunar diameter.

In 1601, he noted that the eye itself possesses an aperture and should be treated in the same way as the aperture in a pinhole camera. Kepler published his theory of vision in 1604.

Until Kepler, the main motivation for studying the nature of light came from a desire to understand vision. René Descartes (1590–1650) appears to be the first person to concern himself with the intrinsic nature of light and the laws of optics. Descartes was a French philosopher and mathematician who had a great impact on western philosophy. He is heralded as the Father of Modern Philosophy. His mathematical contributions included a connection between geometry and algebra that allowed for the solving of geometrical problems using algebraic equations. Descartes promoted the accounting of physical phenomena by way of mechanical explanations. Descartes' main contribution to optics is his book *Dioptrics* that was published in 1637. It deals with many topics relating to the nature of light and the laws of optics. He compares light to a stick that allows a blind person to discern his environment through touch. Descartes used a tennis ball analogy to derive the laws of reflection and refraction of light. The credit for the discovery of the law of refraction is given to Willebrord Snell who derived it using trigonometric methods in 1621. However, Snell did not publish his work in his lifetime. Descartes published the law of refraction 16 years after Snell's death, as Descarte's law of refraction.

Together with Descartes, Pierre de Fermat was one of the two greatest French mathematicians of the first half of the seventeenth century. A lawyer by profession, Fermat made a number of important contributions in analytical geometry, probability, and number theory. He is most well-known for Fermat's Last Theorem (no three positive integers a, b, and c can satisfy the relation $a^n + b^n = c^n$ for any integer n that is larger than 2) that he conjectured in 1621 but could not be proved till 1994. Fermat's major contribution in optics relates to his derivation of Snell's law using the principle of least time. Just as Hero of Alexandria had derived the law of reflection on the basis of the principle of least distance 1400 years before, Fermat argued that light rays going from a point located in a region where it propagates with a particular speed to a point in another region where it propagates with a different speed, it would follow a path that takes the shortest time. This yielded the correct Snell's law.

4.4 Isaac Newton and corpuscular nature of light

Sir Isaac Newton is definitely the defining figure in the history of science. We have seen how he laid down the foundation of classical mechanics. His discovery of the law of gravitation and the laws of motion are among the greatest intellectual leaps in human history. It is, however, interesting to note that the most important experimental contributions to physics made by Newton are all in the field of optics.

He was the first to show that color is a property of light and not of the medium. Through ingenious experiments he could show that the light generated by the sun consisted of all the colors. For example, when the light from the sun passes through a prism, it is dispersed in a rainbow of colors (Fig. 4.2). The red color bends the least and the violet color bends the most. This ability of glass prisms to generate multiple colors has been known since antiquity but it was not attributed to light. Instead, color was considered as a characteristic of the material. What Newton showed was that when a particular color passed through the prism, no such dispersion took place. In a relatively complicated setup, when these colors were combined together and passed through the prism again, Newton recovered white light, proving that white light consisted of all the colors.

The other major contribution of Newton toward optics is his design of the reflecting telescope as shown in Fig. 4.3. All the telescopes through his time were unwieldy refracting telescopes that suffered from chromatic aberrations. The earliest refracting telescope, built in 1608, is credited to Hans Lippershey who got the patent for the design. These refracting telescopes consisted of a convex objective lens and a concave eyepiece. Galileo used this design in 1609. In 1611, Kepler described how a telescope could be made with a convex objective lens and a convex eyepiece lens. Newton designed a reflecting tele-scope where incoming light is reflected by a concave mirror onto a plane mirror that reflected the light to the observer. This design was simple and less susceptible to chromatic aberrations. All the major telescopes that exist today are improved versions of Newton's reflecting telescope.

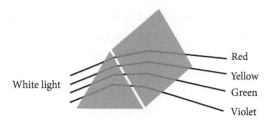

Fig. 4.2 Light from the sun is bent by a prism and is split in different colors. The bending is least for the largest wavelength red light and is the largest for the shortest wavelength violet light.

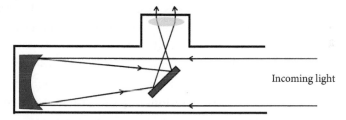

Fig. 4.3 Schematics of the reflecting telescope of Newton. Incoming light is reflected by a concave mirror onto a plane mirror. The reflected light from the plane mirror is sent to the observer.

Newton was also concerned with the nature of light and advocated the corpuscular theory of light. According to him, light is made up of extremely small corpuscles, whereas ordinary matter was made of grosser corpuscles. He speculated that through a kind of alchemical transmutation they change into one another. According to him,

> Are not gross Bodies and Light convertible into one another, ... And may not Bodies receive much of their Activity from the Particles of Light which enter their Composition?

It is surprising that Newton advocated the corpuscular theory of light when there was evidence that supported the wave behavior. For example, Francesco Grimaldi made the first observation of the phenomenon that he called diffraction of light. He showed through experimentation that when light passed through a hole, it did not follow a rectilinear path as would be expected if it consisted of particles but took on the shape of a cone. Newton explained that the phenomenon of diffraction was only a special case of refraction that was caused somehow by the ethereal atmosphere near the surface of the bodies. Newton could explain the phenomenon of reflection with his theory. However, he could only explain refraction by incorrectly assuming that light decelerated upon entering a denser medium because the gravitational pull was stronger.

When Newton was expounding a corpuscular nature of light, his contemporary, Christian Huygens, suggested a wave picture of light. Huygens published his results in his *Traite de la lumiere* (Treatise on light) in 1690. Crucial to his wave theory was the result recently obtained by Olaus Romer (1679) that the speed of light is finite. He considered light waves propagating through the ether just as sound waves propagate through air. He explained the high but finite speed of light by seeking an analogy with the elastic collisions of a succession of spheres that made the ether. The light waves, according to Huygens, were thus longitudinal waves as opposed to the later studies by Fresnel and Maxwell that showed light to consist of transverse waves. Huygens formulated a principle (that now bears his name) which describes wave propagation as the interference of secondary wavelets arising from point sources on the existing wavefront. In propagation each ether particle collides with all the surrounding particles so that "... around each particle there is made a wave of which that particle is the center." This is shown in Fig. 4.4.

4.5   Thomas Young and Young's double-slit experiment

Till the beginning of the nineteenth century, Newton's status was so great, particularly in the British Isles, that few dared to challenge his corpuscular theory of light. It was, however, Thomas Young who, in 1802, conclusively demonstrated the wave nature of light through his double-slit experiment as shown in Fig. 4.5. He described his experiment in these words in *The Course of Lectures on Natural Philosophy and the Mechanical Arts* (1807):

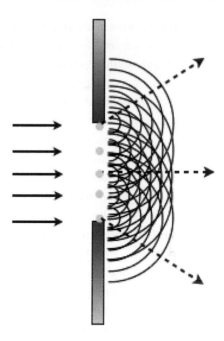

Fig. 4.4 According to Huygens' wave theory, every point on a wave-front is a source of a secondary wave in the forward direction.

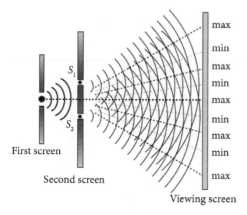

max
min
max
min
max
min
max
min
max

S_1

S_2

First screen

Second screen

Viewing screen

Fig. 4.5 In Young's double-slit experiment, light from a source is incident on a screen with two slits. The light from these slits forms an interference pattern on the viewing screen.

when a beam of homogeneous light falls on a screen in which there are two very small holes or slits, which may be considered as centers of divergence, from whence the light is diffracted in every direction. In this case, when the two newly formed beams are received on a surface placed so as to intercept them, their light is divided by dark stripes into portions nearly equal, but becoming wider as the surface is more remote from the apertures, so as to subtend very nearly equal angles from the apertures at all distances, and wider also in the same proportion as the apertures are closer to each other. The middle of the two portions is always light, and the bright stripes on each side are at such distances, that the light coming to them from one of the apertures, must have passed through a longer space than that which comes from the other, by an interval which is equal to the breadth of one, two, three, or more of the supposed undulations, while the intervening dark spaces correspond to a difference of half a supposed undulation, or one and a half, of two and a half, or more.

With this he firmly established the wave nature of light. Young's double-slit experiment and its implications for quantum mechanics are discussed in Chapter 10.

By repeating his experiment, Young could relate color to wavelength and was able to calculate approximately the wavelengths of the seven colors recognized by Newton that composed white light. According to him "… it appears that the breadth of the undulations constituting the extreme red light must be supposed to be, in air, about one 36 thousandth of an inch, and those of the extreme violet about one 60 thousandth."

The Young's double-slit experiment was not only decisive in debunking Newton's corpuscular theory of light, but it also continued to play a crucial role in our understanding of the nature of light and matter even in the twentieth century. For example, in 2002, *Physics World* published the results of a survey on the all-time Ten Most Beautiful Experiments in Physics. Young's double-slit experiment made not one but two appearances on this prestigious list—at number 1 was the double-slit experiment applied to the interference of electrons that we discuss in Chapter 10 and at number 5 was the original experiment by Young.

Young's double-slit experiment was, however, regarded as highly controversial and counterintuitive in his own time. How can a screen uniformly illuminated by a single aperture develop dark fringes with the introduction of a second aperture? And how could the addition of more light result in less illumination? Young's theory would eventually find broad acceptance, particularly through the works of Fresnel in France.

Augustin Jean Fresnel, a contemporary of Young, championed the wave nature of light based on his work on diffraction. He noted that when light passed through an aperture, one could see a series of dark and bright fringes on the screen as shown in Fig. 4.6. These bright and dark regions extended beyond the geometrical shadow of the region outside the aperture. This diffraction pattern can be explained by the Huygens principle that treats each point on a wavefront as a source of a secondary spherical wave. This leads to the spreading of the wave. Fresnel was able to develop a mathematical theory for these observations based on a wave theory of light and could predict the position of bright and dark lines based on where the vibrations were in phase and out of phase. He published his first paper on the wave theory of diffraction in 1815.

An episode indicates the stunning success of the wave nature of light as formulated by Fresnel. In 1819, Fresnel presented his work on wave theory of diffraction in a competition by the French Academy of Sciences. The committee of judges, headed by Francois Arago, included Jean-Baptiste Biot, Pierre-Simon Laplace, and Simeon-Denis Poisson. They were all prominent advocates of Newton's corpuscular theory and were not well disposed to the wave theory of light. Poisson was, however, impressed by Fresnel's submission and extended his calculations to come up with an interesting consequence:

Let parallel light impinge on an opaque disk, the surrounding being perfectly transparent. The disk casts a shadow—of course—but the

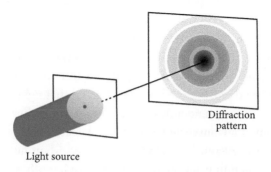

Fig. 4.6 Diffraction from a circular aperture.

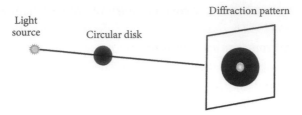

Fig. 4.7 Poisson spot: Light incident on an opaque circular disk forms a bright spot along the axis as a result of Fresnel diffraction.

very center of the shadow will be bright. Succinctly, there is no darkness anywhere along the central perpendicular behind an opaque disk (except immediately behind the disk).

This effect is shown in Fig. 4.7. According to the corpuscular theory, there could be no bright spot behind the disk. As Chair of the Committee, Arago decided to do the experiment and observed the spot himself as predicted by Fresnel's theory. This discovery was an impressive vindication of the wave theory and Fresnel won the competition. This spot is now known as the "Poisson spot" or the "spot of Arago."

Despite the triumph of the wave theory of light, the properties of the polarized light still provided a strong argument in favor of the corpuscular theory, since no explanation from a wave theory had ever been made. Following the success of the wave theory in explaining the interference and diffraction phenomena, Fresnel and Arago embarked upon explaining the properties of the polarized light based on Fresnel's theory. In 1817, Fresnel became the first person to obtain what was later called circularly polarized light. The only hypothesis that could explain the experimental results was that light is a transverse wave. In 1821, Fresnel published a paper in which he claimed that light is a transverse wave. Young had independently reached the same conclusion. The assertion that light is a transverse wave was not readily accepted by many, including Arago. Again, Fresnel was vindicated when he could explain the double refraction from the transverse wave hypothesis. This helped to seal the status of light as a transverse wave.

4.6 James Clerk Maxwell: Electromagnetic waves

It was left to James Clerk Maxwell to complete the classical picture of light as consisting of electric and magnetic waves. This was a truly remarkable outcome of his efforts to unify the two known forces of nature: electric force and magnetic force.

It was known through the work of Michael Faraday that a time rate of change of magnetic field yielded electric force. This discovery had a great impact on our lives as it made it possible to build electric generators, motors, transformers, and electric instruments. These devices make it possible to drive our cars, power our homes, and cook our food. The insight due to Maxwell was that if electricity and magnetism were the two sides of the same coin then a change of electric field should similarly result in a magnetic field. This motivated him to postulate that a time rate of change of the electric field should produce a magnetic field. This realization immediately yielded a wave equation for an electromagnetic wave propagating at the same speed as known for light, 300,000 kilometers per second. A remarkable aspect of this discovery was that the speed of light was a constant expressible in term of some parameters of electricity and magnetism. This would have far-reaching consequences when Albert Einstein laid the foundation of his theory of relativity (Chapter 15) based on this observation.

The picture of light that emerged was thus that of undulations of mutually perpendicular electric and magnetic fields propagating as shown in Fig. 4.8. The direction of the propagation is perpendicular to both the electric and magnetic fields. Maxwell's results were published in 1865. Thus, the light waves were shown to be transverse waves in line with Young and Fresnel as opposed to the picture adopted by Huygens where light was seen as a longitudinal wave propagating through the medium ether. The existence of light as an electromagnetic wave was experimentally demonstrated by Heinrich Hertz in 1888.

The light that we are able to see lies in the visible region of the spectrum. Different colors correspond to different wavelengths. For example, the colors red, orange, yellow, green, blue, and violet have wavelengths 750 nanometers, 625 nanometers, 590 nanometers, 565 nanometers, 485 nanometers, and 450 nanometers, respectively. However, this is not

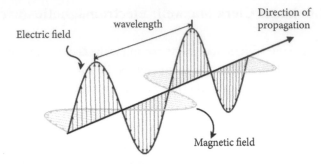

Fig. 4.8 Light of a given wavelength consists of mutually perpendicular oscillating electric and magnetic fields. The direction of propagation of light is perpendicular to both electric and magnetic fields.

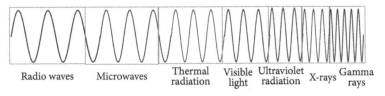

| Radio waves | Microwaves | Thermal radiation | Visible light | Ultraviolet radiation | X-rays | Gamma rays |

Fig. 4.9 The spectrum of electromagnetic waves spread from large-wavelength radio waves to short wavelength gamma rays and beyond.

all. The electromagnetic radiation spectrum is spread over a very wide range as shown in Fig. 4.9. These "colors" lie in the invisible regions. For example, the wavelength of a radio wave that is used in radio transmission is about 1 millimeter, the microwave in the range 1 millimeter to 25 micrometers, the thermal radiation in the range 25 micrometers to 750 nanometers, visible light in the range 750 nanometers (red) to 400 nanometers (violet), ultraviolet in the range 400 nanometers to 1 nanometer, x-rays in the range 1 nanometer to 1 trillionth of a meter, and gamma rays with wavelengths smaller than 1 trillionth of a meter.

As discussed earlier, an important property of the light is the polarization. For an electromagnetic wave, the polarization is represented by the direction of the electric field of the propagating light wave. The concept of polarization will be used to elucidate many aspects of quantum mechanics in Chapter 6.

4.7 Albert Einstein and wave–particle duality

The revival of the particle theory of light, and the beginning of the modern concept of the photon, is due to Albert Einstein. In Chapter 3, we discussed Einstein's 1905 paper on the photoelectric effect. To explain the emission of electrons from a metallic surface irradiated by ultraviolet light rays, Einstein postulated that light comes in discrete bundles, or quanta of energy, in accordance with Planck's hypothesis. This reintroduced the particulate nature of light into physical discourse, not as localization in space in the manner of Newton's corpuscles, but as discreteness in energy.

With the advent of quantum mechanics, the dual nature of light was apparent. There were phenomena such as interference and diffraction that could be explained based on the wave nature of light. Then there were phenomena such as excitation of an atom by absorbing a photon that required a particle nature of light. It was Paul Adrien Dirac who, in a seminal paper published in 1927, synthesized the wave and particle natures of light in a single theory. According to the Maxwell's theory, the light consisted of electromagnetic waves of different frequencies. The oscillating waves could be looked upon as a sort of simple harmonic oscillators. Central to Dirac's quantum theory of radiation was the notion that each color of the electromagnetic waves could be identified as a quantized simple harmonic oscillator.

5

What does an atom look like?

In Science, it is when we take some interest in the great discoverers and their lives that it becomes endurable, and only when we begin to trace the development of ideas that it becomes fascinating.
— *James Clerk Maxwell*

One of the earliest successes of quantum mechanics was an explanation of what an atom looks like. The picture of the atom that emerges is truly startling and almost unbelievable.

An atom is the smallest constituent of an element. The history of the atom is long, going back to antiquity. At the end of the nineteenth century, an atom was considered a tiny particle with no internal structure. The subsequent discovery of electrons, and then the nucleus, provided a view of the internal structure of the atom. The simplest and lightest atom is that of hydrogen. A breakthrough came through the work of Niels Bohr who described the hydrogen atom as consisting of a massive positively charged nucleus with a negatively charged electron circling around it in fixed orbits—something like a tiny planetary system. His work, as pointed out in Chapter 3, represented the third major application of quantum mechanical ideas. For most practical applications, Bohr's model provides an easily comprehensible and visualizable picture.

However, with the advent of quantum mechanics, it was shown that Bohr's model was not true. The quantum mechanical picture is literally "stranger than we can think." The electrons are not circling the atom like the planets around the sun. Instead, they are like fluffy balls surrounding the nucleus. The strangest aspect is that these fluffy balls are not real objects, they represent the probability of finding the electrons when we try to look at them. Thus, the atom consists of nothing "real" but a cloud of probabilities. According to quantum mechanics, electrons are not localized real objects until we try to look at them and then we find them located at some random points around the nucleus. Till then, we cannot associate any reality to the electron. What we have is only a probability of existence at various points. Just imagine that all the objects around us, including ourselves, consist of such atoms piled one on top of the other! Truly bizarre, but the correct picture according to the physical laws as understood now.

Here we present a brief summary of how our understanding of an atom has evolved over a long period.

5.1 From Democritus to John Dalton

The history of the atom starts around 450 BC when a Greek philosopher named Democritus wondered what would happen if an object is cut into smaller and smaller pieces. He thought that a point would be reached where the object could not be cut into still smaller pieces. He called these "uncuttable" pieces "*atomos*." This is where the modern term "atom" comes from. Democritus thought that atoms were infinite in number, uncreated, and eternal, and that the qualities of an object result from the kind of atoms that composed it.

Almost a hundred years later, the Greek philosopher, Aristotle, came up with his own idea of matter which was in contradiction with Democritus' concept of atoms. Aristotle believed that four elements, earth, air, fire, and water, made up everything. For example, a heavy substance such as iron and other metals were made up in large part of the element, earth, and in smaller parts, the other three elements. Similarly, lighter objects could be largely made up of lighter elements, air and fire, and small amounts of heavy elements, earth and water.

Aristotle's influence on our scientific thinking dominated for almost 2000 years. His thoughts about the four constituents of matter were accepted till almost the beginning of the scientific revolution in the seventeenth and eighteenth centuries. By that time, Democritus' ideas were more or less forgotten, but were revived around 1800 by a British chemist, John Dalton. On the basis of his studies on the pressure of gases, he concluded that the gases must consist of tiny particles, atoms, in constant motion. His main interest was in studying the properties of compounds. He concluded that a compound consists of the same elements in the same ratio. Another compound would be made up of different elements in different ratios.

The main points of Dalton's atomic theory can be summarized as follows:

- All elements are made of extremely tiny particles called atoms. Atoms are the smallest particles of matter. They cannot be divided into smaller particles. They also cannot be subdivided, created or destroyed.
- All atoms of the same element are identical in size, mass, and other properties; atoms of different elements differ in size, mass, and other properties.
- Atoms of different elements join together to form compounds. A given compound always consists of the same kinds of atoms in the same ratio.

Many aspects of Dalton's theory were correct and it became a widely accepted theory. However, he was incorrect in assuming that atoms are the smallest particles and are indivisible. Dalton assumed that atoms are like solid spheres. This model had great difficulties in explaining how atoms can be joined together to make compounds. He thought that atoms could have holes and joined together using hooks. This was too simple a model with no experimental support.

Dalton's model was shown to be incorrect when smaller particles like electrons were discovered through the work of J. J. Thomson in 1897, and it was realized that atoms have a much more complicated structure.

5.2 Thomson's model of the atom

Thomson carried out experiments in which he applied a voltage between two metallic plates inside a vacuum tube. He observed that an electric current flowed between the two plates, traveling much further than what we would expect for a current consisting of atom-size particles.

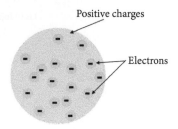

Fig. 5.1 Thomson's plum-pudding model. The atom consists of a sphere of uniform distributed positive charge and negatively charged electrons are embedded like plums in a pudding.

His experiments suggested that the mass of these negatively charged particles should be about 1/1836 times those of a hydrogen atom. He also observed that the mass of these particles was the same regardless of the metal they came from. Thomson had discovered electrons. He also concluded that these particles cannot be the atoms but come from inside the atoms. Electrons are therefore subatomic particles. This was an important discovery.

Next question was how to incorporate the tiny electrons inside the atom. Atoms are electrically neutral, so how could atoms contain negative charges and still be electrically neutral.

Thomson proposed a plum-pudding-type model of an atom in which a spherical atom is like a homogeneously positively charged pudding and electrons are embedded in it like plums. This is shown in Fig. 5.1. This helped to explain the charge neutrality of an atom, Thomson assumed that most of the mass of an atom was due to the positively charged sphere and electrons made only a small contribution.

This was the picture of the atom at the beginning of the twentieth century.

Ernest Rutherford, a physicist from New Zealand, made the next major discovery about the structure of atoms. He discovered the nucleus.

5.3 Rutherford discovers the nucleus

In 1899, Rutherford discovered that certain elements emitted positively charged particles. He called them alpha particles. In 1911, he carried out experiments in which a beam of alpha particles was incident on a very thin sheet of gold (Fig. 5.2). Outside the gold foil he placed an array of

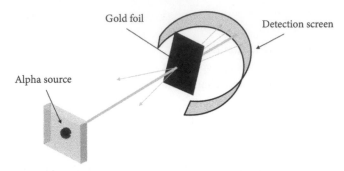

Fig. 5.2 Schematics of Rutherford experiment. A beam of alpha particles is incident on a thin sheet of gold. An array of detectors detects alpha particles after they scatter from the gold foil.

detectors of alpha particles. If Thomson's plum-pudding model was correct then most of the alpha particles would pass through the foil with very small deflection, the deflection caused by a repulsion due to the positively charged "pudding." The experimental results were dramatically different. It was observed that most of the alpha particles passed through the gold foil without any significant deflection. However, a few alpha particles were scattered at very large angles and some even scattered in the backward direction. This was very unexpected. As Rutherford later declared, it was as if a 15-inch naval shell incident on a piece of tissue paper came back and hit you.

The experimental results indicated that the atom was mostly empty space through which the alpha particles passed without any hindrance. But then there were points which sharply repulsed the alpha particles. This clearly showed that Thomson's model of the atom being a sea of positive charge with light electrons embedded in it was incorrect and a new model of atomic structure was required.

Based on the gold foil experiment, Rutherford proposed a new atomic model. His model for an atom was similar to the planetary model as shown in Fig. 5.3. He proposed that most of the mass that carried the positive charge was concentrated in a small area at the center of the atom. He called this area the "nucleus." Negatively charged electrons revolved around the positively charged nucleus like planets revolve around the sun. Thus, most of the atom consisted of almost empty space and almost

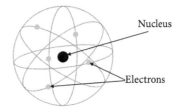

Fig. 5.3 Rutherford model of the atom:
It consists of a massive positively charged
nucleus surrounded by electrons in
random orbits.

all the mass was concentrated in the small nucleus. This model could explain his experiment—the alpha particles could pass through the almost empty space without deflection and a few particles were repulsed by the massive, positively charged nucleus and scattered at very large angles in the backward direction.

Rutherford also showed that the nucleus consisted of protons. These particles are 1836 times heavier than electrons but carry the same but opposite charge as electrons[1]. The number of protons in the nucleus is equal to the number of electrons orbiting the nucleus in random orbits, making an atom electrically neutral. Later it was discovered that, in addition to positively charged protons, there existed electrically neutral particles called neutrons inside the nucleus. The mass of these particles is almost the same as protons. Due to their neutral charge, neutrons were difficult to detect and their experimental observation had to wait till James Chadwick discovered them in 1932.

Rutherford's picture of the atom could explain his gold foil experiments. However, this model was inadequate to explain some other experimental results, most notably the light emitted by various atoms.

Atoms are extremely tiny objects with a typical size of 1 Angstrom or 10^{-10} meter[2]. It is therefore difficult to study directly the properties of atoms. In the nineteenth century, the tools to study the internal structure of the atoms were very limited. The radiation emitted by the atoms provided an important source of information.

---

[1] Mass of the proton is 1.67×10^{-27} kilogram = 0.00000000000000000000000000167 kilogram. Mass of the electron is 9.11×10^{-31} kilogram = 0.000000000000000000000000000000911 kilogram.
[2] 10^{-10} meter = 0.0000000001 meter

5.4 Bohr's model

In 1913, Niels Bohr, a Danish scientist, discovered evidence that the orbits of electrons are located at fixed distances from the nucleus. This was in contrast to Rutherford's atomic model in which electrons orbit the nucleus at random. According to Bohr's model, electrons can exist in well-defined energy levels. These energy levels correspond to orbits of fixed radii. Electrons can only exist in these orbits and not in between. The picture is similar to a ladder where one can stand on one rung or another but not in between the rungs.

An outstanding problem since the late nineteenth century was the emission of certain discrete frequency radiation from hydrogen atoms. The schematics of the experiment that showed the emission spectrum of hydrogen atom is given in Fig. 5.4. Both Thomson's and Rutherford's model of the atom could not explain why light of certain colors or frequencies is emitted and no other frequencies are observed.

Bohr, who had joined Rutherford's research group in 1911, came up with a model of the hydrogen atom that could solve this problem and derive the expression of all the emitted frequencies.

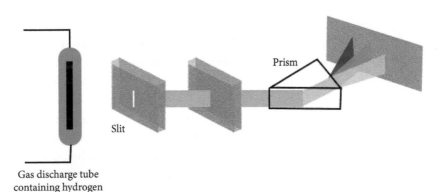

Fig. 5.4 Schematics for observing the emission spectrum of hydrogen. A gas discharge tube containing hydrogen emits radiation which is first collimated by narrow slits and then passed through a prism which deflects light of different colors in different directions.

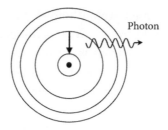

Fig. 5.5 According to Bohr's model of the atom, the electrons exist only in prescribed orbits. Each orbit has a definite energy. When an electron jumps from a higher level to a lower level, a photon of energy equal to the energy difference between the two levels is emitted.

Bohr's model of the atom had the following features. An atom consisted of a positively charged nucleus with electrons revolving around it in fixed orbits. Each orbit corresponds to an energy level. Electrons can exist only in these levels and not in between these levels (Fig. 5.5). The level closest to the nucleus has the least amount of energy. As the radius increases, the energy of the atomic level also increases. Electrons can jump from one energy level to another. When an electron jumps from a higher level to a lower level, it emits a light quantum or a photon whose energy is equal to the energy difference between the two levels. Similarly, an atom absorbs a photon and the electron jumps to a higher level such that the level energy difference is equal to the photon energy.

The challenging task was to develop a theory that could calculate the radii of these energy levels and the energy of electrons in a given level. Most important was that the theory should be able to explain the light frequencies emitted by an atom.

The simplest atom is the hydrogen atom, which consists of a single proton and a single electron. In Bohr's model, the electron revolves around the proton in fixed orbits. Bohr invoked a quantum postulate to find the radii of electron orbits for this simplest of systems. Using this postulate, he could also find the energy of the energy levels in the hydrogen atom. The most remarkable aspect of this model was that it could perfectly explain the spectrum of light emitted by a hydrogen atom.

The picture of the atom that emerged from Bohr's work is as follows. The atom consists of a nucleus consisting of two kinds of particles: positively charged protons and electrically neutral neutrons. The negatively charged electrons move in well-defined orbits around the nucleus. The number of electrons is equal to the number of protons—making the atom electrically neutral. The protons and neutrons are over 1800 times heavier than the electrons. Thus, most of the mass of the atom is concentrated in the nucleus.

The number of protons in the nucleus determines the element—hydrogen atom has single proton, helium has two protons, lithium has three protons ... oxygen has eight protons, and so on. The number of neutrons does not affect the chemical properties of the atom. The simple hydrogen atom has a single proton and no neutron. But another kind of hydrogen, called deuterium, has one proton and one neutron in the nucleus. A helium atom has two protons and two neutrons in the nucleus. Again, another kind of helium has two protons and one neutron. An oxygen atom has eight protons and eight neutrons in the nucleus, nitrogen has seven protons and seven neutrons in the nucleus.

In any atom, the number of orbiting electrons is the same as the number of protons in the nucleus. As mentioned above, these electrons revolve around the nucleus in fixed orbits. The energy of these orbits is "quantized." For example, if the energy of the lowest level is E, then the energy in the next higher allowed orbit is four times E, i.e., $4E$. The energy of the electron in the next higher allowed energy level is nine times the energy of the lowest level, i.e., $9E$. And so on. No electron can exist in between these fixed orbits. In any given level, the maximum number of electrons is also fixed. For example, a maximum of two electrons are allowed in the allowed orbit closest to the nucleus with minimum energy. In the next orbit, a maximum of eight electrons are allowed. Therefore, hydrogen has its only electron in the orbit with the minimum energy and oxygen has two electrons in the lowest energy orbit and the remaining six electrons in the next higher orbit. When the electron in an outer orbit jumps to a lower orbit, a photon is emitted whose energy is equal to the energy difference between the two orbits.

Bohr's success in proposing a model that could explain the radiation emitted by a hydrogen atom was a major vindication of the quantization postulate of Planck and Einstein. However, the model was built on certain assumptions based on similar quantum conditions that were ad hoc and were not rooted within a formal theory. Bohr's model could explain the light emitted by hydrogen only. A similar postulate could not explain the radiation emitted by the next simplest atom, the helium atom.

There was another serious flaw in Bohr's model. It had been known for a long time that an orbiting charged particle radiates energy. Thus, in Bohr's model, an orbiting electron should emit radiation. Radiation carries energy. Therefore, electrons should lose energy and, as a result, spiral down and eventually collapse onto the nucleus. Thus, the atom, according to Bohr's model, should not be stable. If this were true then there would be no atoms. Life as we know it would not exist.

In spite of these serious issues, Bohr's model that explained the radiation emitted by hydrogen atoms was perceived as a major success and Niels Bohr was awarded the 1922 Nobel Prize. It is ironical that Bohr's work would become obsolete within three years of his winning the Nobel Prize when a new theory, the quantum theory, would emerge in 1925. Bohr's model has, however, stayed with us mainly because it is easy to visualize.

The true picture of the atom emerged when quantum mechanics was developed. The emerging picture is truly bizarre and difficult to comprehend.

5.5 The quantum mechanical picture

The results of Planck, Einstein, and Bohr gradually created a realization that Newtonian mechanics must be an inadequate theory to explain phenomena at the atomic level. The period between 1913 and 1925 was a period when new phenomena were being discovered that could not be explained by classical theory. The foundations of Newtonian mechanics were crumbling and the need for a new theory was being felt very urgently. The breakthrough came in 1925/26 when quantum mechanics was born through the works of Heisenberg, Schrödinger, Born, Dirac, and others.

One big test of the new theory was to solve the problem of the hydrogen atom, not through a postulate as Bohr did, but as a result of a formal theory—a theory that could be applied equally well to essentially all the problems of physics even at the level of our daily experience.

Quantum theory showed that, contrary to the Bohr's model of atom, electrons do not travel in fixed orbits. In fact, each electron with a certain amount of energy is described by a "wave function." We explain the physical nature of the wave function in Chapters 7 and 8. Here we present a simple pictorial view of the atom.

In Fig. 5.6a, a picture of the atom is shown when it is in the lowest energy. The electron is no longer described by an orbit. Instead, it appears smeared around the nucleus. The correct interpretation is that

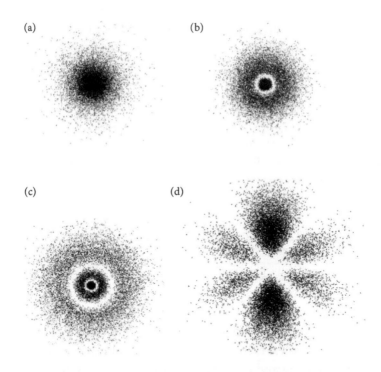

Fig. 5.6 The pictures of hydrogen atom in different energy states. The clouds represent the probability of finding the electron around the nucleus that is not depicted here. The dense areas represent where the probability of finding the electron is large. The lowest energy state is depicted in (a), the first excited state in (b), the third excited state in (c), and the fourth excited state in (d).

the cloud around the nucleus represents the probability distribution of finding the electron. The probability of finding the electrons is higher in the denser regions. This picture has no resemblance to anything we observe in the world around us. In Fig. 5.6b, the probability distribution is given for the electron in the first higher energy level.

There are several counterintuitive aspects with this picture of an atom. First, the probability distribution peaks at some points where the probability of detecting the electron is maximum. However, it is non-vanishing far, very far, away from the atom. There the probability becomes vanishingly small but not zero. This means that there exists the possibility of finding the electron at large distances, one millimeter, one meter, or even one kilometer away. Secondly, for different energies, the electron has overlapping probability distributions. Thus, if, after the measurement, the electron is found at a particular location, we cannot infer the energy of the electron.

Another strange feature of an atom is how empty it is, that is if we think of it with our classical intuition. The atom is usually spherical in shape but not always. The size of the atom is about the tenth of a billionth of a meter. This means that a row of 100 million atoms placed next to each other would stretch to a centimeter. As mentioned above, the central part of the atom is the nucleus consisting of protons and neutrons. The nucleus contains more than 99.9 percent of the total mass whereas the remaining mass is carried by the electrons. However, the size of the nucleus is 1/100,000 the size of the atom.

Just to acquire a perspective, if the nucleus is the size of a tennis ball, the atom would be the size of a small town (about 7 kilometers in diameter) and an electron would be the size of a pea. The rest is empty space. It is truly amazing that, contrary to our impression of an atom of being a solid sphere, it is almost entirely empty—just like a pea and a tennis ball in a sphere of 7 kilometers diameter. Here the numbers we have given correspond to a hydrogen atom, but they are not much different for heavier atoms. It is hard to believe that all the objects around us are made up of atoms that are so hollow.

However, we should be careful with this picture. As mentioned above, the true quantum mechanical picture is that of a nucleus surrounded by a cloud. But this cloud does not represent a smeared-out electron. Instead, this cloud represents the probability distribution of finding the electron. The atom is mostly empty yet occupied with a cloud of probability distribution. Mind-boggling indeed!!

5.6 The nucleus and the forces of nature

So far, we have talked about the electrons surrounding the nucleus. But what about the nucleus itself? There are a number of fundamental questions that need to be addressed.

First, we note that the nucleus consists of protons and neutrons. A question of interest is whether these particles are the most elementary particles with no smaller constituents or whether they are formed by a combination of even more elementary particles.

Second is the phenomenon of radioactivity, where some nuclei are able to emit three kinds of particles, a helium nucleus consisting of two protons and two neutrons called an alpha particle, an electron called a beta particle, and a photon called gamma radiation. These emissions take place in heavier nuclei and happen completely randomly. How does this happen? What kind of forces lead to the emission of these particles?

Third, we note that the number of protons and neutrons inside the heavier nuclei can be large. For example, there are 82 protons and 126 neutrons in the nucleus of an atom of lead. The radius of the nucleus is 5.56 millionth of a nanometer. Similarly, a uranium atom consists of 92 protons and 126 neutrons compactly confined in a sphere of radius 11.7 millionth of a nanometer. It is known that like charges (both positive or both negative) repel each other and the repulsive force becomes very large when these charges are very close to each other. How can a nucleus be so stable holding a large number of positively charged protons in a very compact volume?

And fourth, there are fusion and fission processes that are the source of immense energy. In the fusion process, hydrogen nuclei combine together to form a helium nucleus and, in the fission process, a very heavy nucleus like that of uranium, when hit by an energetic particle, is split into two. In both processes, an enormous amount of energy is released that can be used both for constructive and destructive purposes. The fusion process is responsible for the hydrogen bomb and a tremendous effort continues to harness this energy for peaceful purposes. The fission process is the source of energy via nuclear reactors and is the process that takes place in an atomic bomb.

In the following, answers to these questions are given based on the forces of nature and their characteristics.

It was discovered in 1964, independently by Murray Gell-Mann and by Georg Zweig, that the nuclear particles, protons and neutrons, are themselves composed of elementary particles called quarks. These quarks come in six different varieties or "flavors": up, down, strange, charm, bottom and top (in ascending order by mass). These colorful names are chosen for quarks with different physical characteristics. There is nothing strange about the strange quark and there is nothing charming about the charm quark etc. The quarks can combine together to form different sub-atomic particles. For example, a proton is made of two up and one down quarks and a neutron is made of an up and two down quarks as shown in Fig. 5.7. A change of flavor, an up to down, can convert a proton into a neutron. Unlike protons and neutrons, the quarks are never found individually. The quarks that form protons and neutrons hold themselves so tightly that it would require an extremely large force to separate them.

Next, we address the questions about what kind of mechanisms lead to the nuclear processes like radioactivity and the confinement of quarks in protons and neutrons that keeps the nucleus stable. This brings us to the forces of nature that keep the entire universe in equilibrium and explains all the phenomena that happen within the atom, the nucleus and outside.

There are four forces of nature.

The most visible is the force of gravitation. When an object is released, it falls to earth. Similarly, the planets circle the sun and the moon circles

(a)   (b)

Fig. 5.7 (a) A proton consists of two up and one down quarks, and (b) a neutron consists of two down and one up quarks.

Proton   Neutron

the earth. All these are consequences of the force of gravitation. The force of gravitation is mostly understood in terms of the Newtonian law of gravitation. As discussed in Chapter 2, the force of gravitation between two objects is an attractive force. It is proportional to the masses of the objects and inversely proportional to the distance between them. The force of gravitation is long-range—it extends to infinite distances gradually decreasing as the distance between the objects increases. With the advent of Einstein's general theory of relativity, our understanding of gravitation underwent a revolutionary change. Instead of a force, it is understood as a manifestation of the curving or warping of space and time in the presence of massive objects. This amazing picture will be discussed in detail in Chapter 16. An unresolved problem is how to formulate the quantum theory of gravitation.

The other three forces, namely the electromagnetic force, the weak nuclear force, and the strong nuclear force, can be defined within a unified framework. As we see below, the weak and the strong forces play crucial roles in keeping the nucleus stable. Within the framework of a quantum theory, these forces are based upon exchange of particles.

5.7 Electromagnetic force

First let us consider the electromagnetic force. As discussed above, all atoms consist of two types of charges, negatively charged electrons and positively charged protons. An electric force exists between charges—the same charges, such as two electrons or two protons, repel and the opposite charges attract each other. When the electric charges move, they produce magnetism. Unlike electric charges, there are no magnetic monopoles (only a north pole or only a south pole). The two poles always come together. If a magnetic bar with a north and a south pole is split into two, two magnets are obtained, each having both a north and a south pole. The same magnetic poles repel and the opposite poles attract each other. Till the middle of the eighteenth century, the electric and magnetic forces were considered independent. It was James Clerk Maxwell who unified the two forces in 1873 and showed they are the two sides of the same coin.

With the advent of quantum theory, an important question was how to formulate electromagnetic force. For example, what is the actual quantum process that is responsible for the repulsion of two electrons or two protons. The picture that emerged through the work of Richard Feynman, Julian Schwinger, and Shin'ichiro Tomonaga in 1947 was that these forces are mediated by the electromagnetic particles called photons. They were awarded the Nobel Prize for Physics in 1965 for this seminal work.

To illustrate this point, we consider two electrons approaching each other. When they come close enough, the repulsion force between them becomes so strong that they have to follow a trajectory that takes them farther away. But how can this happen? Quantum mechanics explains this "scattering" process by assuming that a photon is emitted by one electron and it is absorbed by the other electron. The mathematical details of this process are quite complicated. However, a simple way of understanding the scattering process is to realize that the electron that emits the photon and the electron that absorbs it both experience a "recoil" force changing their trajectory. For example, when the electron emits a photon in the forward direction, the electron recoils in the backward direction. This is similar to a rifle recoiling in the backward direction when a bullet is fired in the forward direction. Similarly, when the second electron absorbs the photon, it moves in the forward direction. These photons cannot be observed and are called *virtual* photons.

This process, and many similar processes, are very conveniently described by Feynman diagrams. For electron–electron scattering, the Feynman diagram is shown in Fig. 5.8. Here time moves from down

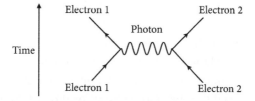

Fig. 5.8 Feynman diagram for electron–electron scattering. The vertical axis represents the time. Two electrons approaching each other are repelled via an exchange of a photon.

to up. We notice two electrons approaching toward each other. This is represented by two arrows pointing upward. When they are close enough, they exchange a *virtual* photon which is represented by a wavy line. As a result, the paths of the incoming electrons change directions and they get farther away from each other representing the repulsion between two electrons. An advantage of using Feynman diagrams is that the strength of the force can be calculated in a series of steps which assign contributions to each of the types of Feynman diagrams associated with the force.

With this background, we can discuss the two nuclear forces, the weak and the strong force.

5.8   Weak nuclear force

The weak force is stronger compared to gravity but is weaker than the electromagnetic and the strong forces. The weak force is an attractive force that acts at very small distances between particles that are closer than 10^{-16} meter, a distance one-tenth the size of a proton. At larger distances, the weak force decreases and approaches zero.

The weak force was first introduced in 1933 by Italian physicist, Enrico Fermi, to explain beta decay which corresponds to the emission of an electron from the nucleus. During a beta decay, a neutron inside the nucleus changes into a proton and expels an electron and a massless particle called a neutrino. The expelled electron is called the beta particle. The net result is that the number of protons inside the nucleus increases by one, thus changing the nature of the element. This happens because each distinct element is characterized by the number of protons in the nucleus. For example, beta decay in a carbon atom which has 6 protons in its nucleus converts it into a nitrogen atom with 7 protons.

In the 1960s, Sheldon Glashow, Abdus Salam, and Steven Weinberg unified the electromagnetic force and the weak force into an electroweak force. They received the 1979 Nobel Prize for Physics for this important work. They showed that, just like the electromagnetic force is carried by an exchange of photons, the weak force arises from an exchange of particles called W and Z bosons. The W bosons are charged particles, with the positively charged particles labelled as W^+ and the negatively charged as W^-. The Z bosons do not carry any charge.

The unified electroweak theory required all the particles—photons, W, and Z bosons—to be massless. A massless exchange particle meant a long-range force like the electromagnetic force. However, the weak force is known to be extremely short-range. This behavior of the weak force required the W and Z bosons to carry a mass larger than that of a proton. In a series of three papers in 1964, by Robert Brout and François Englert, by Peter Higgs, and by Gerald Guralnik, Richard Hagen, and Tom Kibble, a process now known as the Higgs mechanism was proposed to assign masses to the W and Z bosons. The W bosons were discovered in 2012 at the CERN particle accelerator. In 2013, Higgs and Englert were awarded the Nobel Prize for Physics.

As mentioned above, the weak force plays an important role in beta decay when an electron is emitted by the nucleus. In this process, a neutron is converted into a proton and an electron. The Feynman diagram for beta decay is shown in Fig. 5.9. The W boson changes the make-up of the particles. By emitting a W boson, the weak force changes the flavor of a quark. In Fig. 5.9, a down quark d is converted into an up quark u via an exchange of the negatively charged W^-, changing a neutron into a proton. In the process, an electron and a particle called an anti-neutrino are also emitted. Similarly, an exchange of W^+ can convert an up quark u into a down quark d, thus changing a proton into a neutron. This process plays a key role in triggering nuclear fusion that is responsible for the burning of a star like the sun and generating heat and light. This process is discussed in Chapter 14.

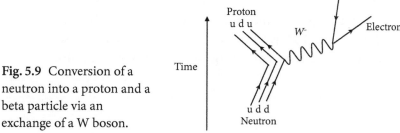

Fig. 5.9 Conversion of a neutron into a proton and a beta particle via an exchange of a W boson.

5.9 Strong nuclear force

The force that provides stability to nuclear particles, protons and neutrons, is the strongest of all the forces, called the strong nuclear force. The strong force is carried by a particle called a gluon. This particle helps to "glue" together the quarks inside a proton and a neutron. The strength of the attractive strong force is such that it is not possible to isolate a quark using the kind of energies available in any laboratory. The strong force is also responsible for the attractive force between the protons keeping the nucleus stable despite the repulsive electromagnetic force between protons.

A remarkable fact about the strong force is that, unlike the other three forces of nature, it does not decrease with the distance between the two particles. Instead, it becomes stronger as the particles separate up to a distance of the size of a proton. The situation is similar to a mechanical spring in which the force to bring the two ends closer becomes stronger the more we stretch the spring. However, there is a limit to how far the spring can be stretched.

In addition to keeping three quarks together inside a proton or a neutron, there is a residual force that acts outside the proton and neutron. This residual force is responsible for keeping the protons and neutrons together inside the nucleus by overcoming the repulsive electromagnetic forces between protons. When the number of protons increases, the size of the nucleus becomes bigger. The electromagnetic forces between distant protons start becoming more dominant than the strong force. Ultimately, the nucleus can be broken by a high-energy particle hitting the nucleus via the process called fission. During nuclear fission, high-energy particles and gamma rays are emitted that can cause other nuclei to undergo the fission process, thus inducing a *chain reaction*. Energy from the fission of heavy nuclei, such as uranium-235 and plutonium-239, is what powers nuclear reactors and is the source of the immense destructive energy released in atomic bombs.

PART 2
MYSTERIES OF QUANTUM MECHANICS

6

A single photon—An amazing quantum system

I therefore take the liberty of proposing for this hypothetical new atom, which is not light but plays an essential part in every process of radiation, the name photon.

— *Gilbert N. Lewis*

Quantum mechanics has many aspects that are highly counterintuitive. It would help to consider a simple system that can bring out these amazing, and sometime mind-boggling, features in as transparent manner as possible. In this chapter, we consider such an example—a single photon that is polarized in a certain direction. Even this description "single photon that is polarized in a certain direction" is too complicated and remote for someone not exposed to advanced level physics. We therefore explain this notion by first examining a system that is quite easy to visualize: an object like a needle passing through a screen. Once this example is understood, it should be possible to consider the realistic system of a photon without knowing in detail about its nature. The objective of this chapter is to convey the features of one of the simplest quantum systems before embarking on the description of the mysterious consequences of quantum theory.

6.1 A heuristic example: "Quantum needle"

At the outset, it should be stated that the "quantum needle" described in this and later chapters is a fictitious object that does not exist in the real world. It is used, for a layman, to understand the intricacies and mysteries of the quantum world. Once we understand the behavior of this fictitious object, it should become straightforward to understand the completely analogous objects like a polarized photon and the spin of an electron that are readily used in experiments in physics laboratories.

The system we consider is a long object, like a needle, going through a screen with two doors, one in the horizontal direction and another in the vertical direction as shown in Fig. 6.1. It is obvious that if the needle is moving in the upright position, it will pass through the screen via the vertical door and if, it is moving in the lying-down position, it will go through the horizontal door. An important point is that, if we did not know the needle's orientation before passing the screen and, after passing through the screen, we find it in the horizontal position, we will conclude that it was in the horizontal position even before passing through the screen. Same is true if we find the needle in the vertical position after the screen. Trivial!!

Next, let us consider the same needle passing through the screen which is rotated by 45 degrees. In this case, there are again two doors, which, instead of being in the horizontal and vertical directions, are oriented at 45 degrees and 135 degrees. This is shown in Fig. 6.2. In such a situation, the needle will hit the screen and will be unable to go through in both situations—when it is moving in the upright position as well as when it is in the lying-down position. So far there is no mystery—both situations, unrotated and rotated screens, can be understood with our conventional intuition.

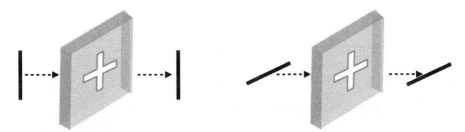

Fig. 6.1 A needle oriented in the vertical direction passes through the screen in the vertical direction and a needle oriented in the horizontal direction passes in the horizontal direction.

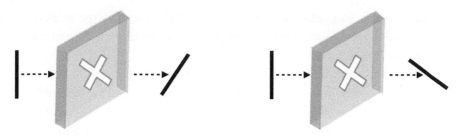

Fig. 6.2 The doors are rotated by 45 degrees. A "quantum" needle in the vertical direction has equal probability to pass through the doors oriented at 45 degrees and 135 degrees.

Now comes the strange behavior!!

For such a fictitious needle, a restriction is imposed—the needle has to pass through the screen no matter what. Hitting the screen and consequently reflecting from the screen is not an option. As we see in the next section, such behavior is exhibited by a quantum object. Therefore, we call such a fictitious needle a quantum needle.

For such a quantum needle, the first part remains the same—if the needle is moving in the upright position, it will pass through the screen via the vertical door and if it is moving in the lying-down position, it will go through the horizontal door.

But what happens when the screen is rotated by 45 degrees? What happens if the quantum needle, moving in the vertical position, arrives at the rotated screen? It again sees two doors. However, instead of being in the horizontal and the vertical directions, these are oriented at 45 degrees and 135 degrees. As the quantum needle can no longer keep moving in the vertical position, it will have two choices: either pass through the door oriented at 45 degrees or pass through the door oriented at 135 degrees. There is no preference that guides the needle which door to pick. The result is that it can go through either door. In such a case, there is a 50 percent probability that it will pass through the door oriented at 45 degrees and a 50 percent probability that it will pass through the door oriented at 135 degrees. The same situation will arise if the needle approaches the screen oriented in the horizontal position. Again, there will be a 50 percent probability that it will pass through the door oriented at 45 degrees and 50 percent probability that it will pass through the door oriented at 135 degrees. If there are a large number of needles with identical orientation (vertical or horizontal), 50 percent will pass through each door.

For such a quantum needle, no matter how much we rotate the screen (5 degrees, 10 degrees, ⋯ 70 degrees, or 80 degrees), it will pass through, the only difference will be the probability. As an example, let us consider the screen that is rotated by (say) 10 degrees. In this case, the horizontal door will be at 10 degrees and the vertical door at 100 degrees. An upright needle will pass through the door at 10 degrees with a 3 percent probability and the door at 100 degrees with 97 percent probability. These probabilities will change to 12 percent and 88 percent for the doors oriented at 20 degrees and 110 degrees, respectively.

The analysis will be the same if, instead of rotating the screen, we rotate the needle. Thus, as an example, the needle rotated at 45 degrees, when passes through the screen with doors in the vertical and horizontal directions, will have 50 percent probability each to pass through either the vertical door or the horizontal door.

This is quite mysterious and strange. The most amazing part is that the orientation of the needle does not seem to have any objectivity. The orientation of the doors in the screen, acting as a measurement device, determines the orientation of the needle. Does it then even make sense to talk about the orientation of the needle without making a measurement? This simple, but deep, question was behind the major philosophical arguments between the two giants of physics of the twentieth century, Albert Einstein and Niels Bohr. This question continues to be at the foundation of any discussion regarding *reality*. More about it in Chapter 12.

If such a needle exists, what conclusions can be drawn? Amazingly, many of the quantum mechanical mysteries can be explained and understood through this simple but apparently unrealistic set-up. But first, what is an example of a quantum needle—a needle that shows explicitly the behavior discussed above?

As stated earlier, we do not encounter such a needle in real life. This is true because a big object like a needle is normally adequately described by Newtonian laws and quantum behavior of such objects can be manifested only under conditions which will be the subject of later chapters. Quantum behavior is exhibited by objects that are small, of the size of an atom or even smaller.

Here we present an extremely simple example: a single photon with the property of polarization. The direction of polarization of a single photon is like the orientation of a quantum needle and a polarization beam splitter corresponds to the screen with two mutually perpendicular doors. Let us first understand this object before discussing the startling conclusions associated with the quantum mechanical predictions.

6.2 Polarized photon

Light consists of waves—the ripples consisting of alternate maxima and minima, like ocean waves. In general, light consists of different colors, each color corresponding to a wave of somewhat different characteristics. An important property of light waves is polarization. The polarization of a wave is described as the direction of the ripples which is perpendicular to the direction of the propagation of the wave. The direction of polarization can therefore be in all possible directions perpendicular to the direction of propagation.

Natural light sources such as sunlight as well as many artificial sources such as light bulbs emit light with random directions of polarization. Such a light beam is described as unpolarized as it lacks a certain well-defined direction of polarization. When such a light passes through a material called a polarizer, the filtered light oscillates in the well-defined direction depending on the orientation of the polarizer and the light becomes polarized in a certain direction (Fig. 6.3).

Just as material objects are made up of atoms, light can be treated as consisting of photons. A single photon can be the smallest amount of light of a certain color. A property of the photon is its polarization. The direction of the polarization of a single photon is clearly analogous to the orientation of the needle—the quantum needle.

The role of the screen, with two mutually perpendicular doors, is played by a polarizing beam splitter which is a device, usually a crystal, that allows a photon polarized along a certain direction to pass through. Thus, a photon polarized in vertical direction can go through the

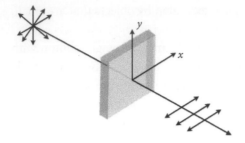

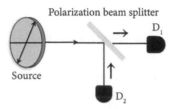

Fig. 6.3 Unpolarized light passing through a polarizer with its polarizing axis along the horizontal direction.

Fig. 6.4 If a photon is polarized in a direction making an angle 45 degrees, ↗, with the polarization axis incident on a polarization beam splitter (PBS), it can pass through as a photon with horizontal polarization represented by the state → in the forward direction or get reflected with vertical polarization represented by state ↑ in the downward direction.

polarizing beam splitter undeflected and a photon polarized in the horizontal direction is deflected in a direction perpendicular to the direction of the propagation of the photon. Just like the quantum needle, a photon polarized in the vertical or horizontal directions, when passing through a polarizing beam splitter rotated by 45 degrees will have equal probability to go through polarized along 45 degrees or 135 degrees (Fig. 6.4).

Next, we introduce some notations and consider some special cases.

We designate the 0 degree (horizontal) and 90 degree (vertical) orientations of the polarization by → and ↑ and the 45 degree and 135 degree orientations by ↗ and ↖. Correspondingly the polarization beam splitter that can select the 0 degree and 90 degree orientations of the polarization is represented by ⊕ whereas the polarization beam splitter that can select the 45 degrees and 135 degree orientations is represented by ⊗. This makes the analysis simple and visually clearer. For example, the statement that a vertically polarized photon in state ↑, when passing through a polarizing beam splitter rotated by 45 degrees, ⊗, has equal probability to pass through in states ↗ and ↖ makes sense. Similarly, it becomes easy to argue that a photon polarized at 45 degrees in state ↗, when passing through a polarizing beam splitter with orientation ⊕ has equal probability to pass through in states → and ↑.

All the possibilities of the polarization of the incident photon and the orientations of the beam splitter are given in Table 6.1.

The philosophical consequences of this behavior of the photon are profound. This simple system clearly demonstrates that the state of the polarization of the photon depends on the setting of the measuring apparatus, the orientation of the polarizing beam splitter in this example. The dependence of the experimental outcome of the same object on the orientation of the apparatus is a very important difference between the usual classical description and the quantum description.

This raises an important question: Can we objectively define the state of the polarization of a single photon? As stated above, this question lies at the heart of the conceptual foundations of quantum mechanics. In the light of the above discussion, a question such as "What is the state of polarization of the given photon?" is ambiguous and incomplete. A well-defined question would include the description of the experimental apparatus (in this case, the orientation of the polarization beam splitter as the measurement device) as well. So, a complete question should be: "What is the state of polarization of the given photon if it passes

Table 6.1 The output polarization states of the photon are given for all possible orientations of the incident photon and the polarization beam splitter.

Polarization of incident photon	Orientation of beam splitter	Polarization of the output photon
↑	⊕	↑
↑	⊗	↖ or ↗
→	⊕	→
→	⊗	↖ or ↗
↗	⊕	↑ or →
↗	⊗	↗
↖	⊕	↑ or →
↖	⊗	↖

through a polarization beam splitter oriented in a certain direction?" We can further illustrate the role of measurement in defining a physical property such as polarization by the following discussion.

Consider a single photon with polarization ↑. What happens when this photon passes through a polarization beam splitter with orientation ⊗? As seen above, the system will be found in the state ↗ with a 50 percent probability or in the state ↖ also with 50 percent probability. Let us assume that the photon is found in the polarization state ↗ If, after this measurement, we measure the polarization again in the original orientation ⊕, our classical intuition tells us that the photon should be found in the state ↑ with unit probability, that is with certainty. However, this is not what happens. The polarization state ↗, when passing through the polarizer with orientation ⊕, yields the outcomes ↑ and → with equal probability. Thus, there is a 50 percent chance that the photon will be found in the horizontally polarized state →. This is a counterintuitive result that could not be expected in classical mechanics.

This result motivates us to ask a related question: "Can we associate two polarization components ↑ and ↗ for the same photon?" The answer is an emphatic No!

The above analysis provides a beautiful and simple example of Bohr's principle of complementarity, namely, two observables are *complementary* if precise knowledge of one of them implies that all possible outcomes of measuring the other one are equally probable. In the above example, if we measure the polarization in the polarizer orientation ⊕ and the outcome is ↑, then subsequently the outcome with the polarizer orientation ⊗ becomes completely uncertain, with 50 percent probability each for the outcomes ↗ and ↖. Measurement disturbs the system.

This result provides the foundation for some of the most dramatic successes in the field of secure communication as discussed in the next section.

6.3 Quantum cryptography

In summer 2018, a research paper appeared in the prestigious physics journal, *Physical Review Letters*, which rattled the political establishments in capitals all over the world. It is very rare that a single research paper has such an impact. The research paper reported the results of an experiment that was done in China. A team led by Jian-Wei Pan demonstrated a scheme for perfectly secure communication at a distance of 2500 kilometers within China (from Xinglong to Nanshan) and at a distance of 7600 kilometers between China and Austria (between Xinglong and Graz) via a satellite named after the ancient Chinese optical scientist, Micius (Fig. 6.5). The exchanged message was in the form of pictures. For example, Chinese scientists sent the picture of their great hero, Micius, and Austrians reciprocated by sending the picture of their big star, Erwin Schrödinger, whose name will reverberate throughout this book. These pictures were exchanged in a manner that no one, besides the sender and receiver, could find what information, message, or picture was exchanged between them. This was an example of the perfect secrecy of the communication over channels that were accessible to everyone.

This experiment demonstrated the supremacy of China in the field of quantum communication. Immediately after the experiment was reported, President Trump announced a research initiative of one billion dollars in the USA with the explicit objective of attaining parity with Chinese scientists in quantum technology. Similar initiatives were launched in many other countries as well as the European Union.

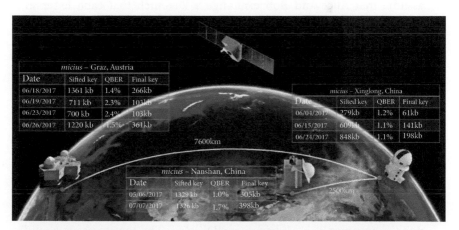

Fig. 6.5 The implementation of a secure quantum communication protocol via satellite. (Reprinted from S.-K. Liao et al., *Physical Review Letters* 120, 030501 (2018) with permission of the American Physical Society).

What is really remarkable about this highly consequential experiment is that it can be fully understood by knowing just how a polarized photon passes through a polarizing beam splitter. Here we show this explicitly, providing a beautiful, yet exciting, application of the novel features of quantum mechanics. But first, a little background!

Since ancient times, a topic of great interest has been the exchange of information over long distances with complete secrecy and security. The sender of the information should be confident that her message is received by the receiver in such a way that no one else has access to this message. This topic has many applications ranging from commercial transactions where the information transferred is kept secret from a potential eavesdropper to military applications where the security and confidentiality of information can make the difference between victory and defeat.

Cryptography is a method of secure communication between two or more parties. Here, the sender, we call her Alice, and receiver, called Bob, first exchange a key in a secure manner. Alice then encodes her message with the key, which is known only to Alice and Bob. The message gets encoded in a way that makes it scrambled. This encoded message is then sent to Bob who can decode the message using the key.

As an example of this kind of old-fashioned cryptography, let us assume that Alice and Bob exchange a key such that each letter in alphabet is shifted by one, i.e.,

$$A \to Z, B \to A, C \to B \cdots \cdots Z \to Y$$

This is perhaps the simplest of the keys that we use for the purpose of illustration. Typically, keys are much more complex. Let us suppose that Alice wants to send the message:

I AM HAPPY TO BE READING THIS BOOK

The message is encoded with the above key by displacing each letter by one ('I' becomes 'J', 'AM' becomes 'BN' etc.). The sent message would therefore be

J BN IBQQZ UP CF SFBEJOH UIJT CPPL

The message is then sent through a public channel like a telephone line or via the internet. If, during transmission, an eavesdropper, Eve, is able to intercept this message, it would not make any sense to her unless she knows the key. When the message reaches Bob, he can decode the message by shifting each letter by one in the opposite direction ('J' becomes 'I', 'BN' becomes 'AM' etc.) and the full message is recovered.

There are, however, two problems with this type of cryptography. First Alice and Bob should exchange the key through highly reliable and secure channels. For example, such a key cannot be exchanged on a telephone line that can be intercepted rather easily by a resourceful eavesdropper. A secure channel could be through a physical contact between Alice and Bob or their reliable representatives who can then travel to their destinations before starting the exchange of information. The second problem is that a clever eavesdropper, who has complete access to the public channel through which the message is being sent, can, by a careful analysis of the sent information, reconstruct the key. Both Alice and Bob may continue communicating without knowing that their key is known to a clever Eve and consequently their communication security is compromised. Of course, many clever schemes for conventional cryptography have been proposed and implemented, but ultimately all of them are plagued with these problems.

The quantum ideas developed above can amazingly be used to overcome both these problems: The key is exchanged on a public channel like a fiber-optics based telephone line and, if an eavesdropper attempts to intercept, Alice and Bob can find out that their communication is being intercepted. This is almost like magic. But quantum mechanics can accomplish these tasks as was so dramatically demonstrated in the Chinese satellite experiment mentioned above.

We note that, in present-day electronic communication, the information and the messages are not exchanged in the old-fashioned way, through letters. Instead, they are transmitted via a sequence of 0's and 1's. Each letter is designated by a string of 0's and 1's. For example, the letter A is represented by 0100 0001, B by 0100 0010 and so on. So, the word 'Book' corresponds to the sequence

$$B \qquad o \qquad o \qquad k$$
$$0100001001101111011010111101101011$$

This example shows that any message or information can be communicated through binary numbers "0" and "1." Thus, a secure way of sending numbers would ensure that any message can be sent in a secure manner.

Before discussing quantum cryptography, we discuss the procedure for encoding and decoding the message via a random key made up also of binary numbers, schematically shown in Fig. 6.6.

Typically, the transmission via a random key takes place as follows: The objective is to send the data (a sequence of bits 1 and 0). As an example, we consider the message:

$$\textit{Message}:\ 1\ 1\ 0\ 1\ 0\ 0\ 0\ 1\ 1\ 1\ 1\ 0\ 1\ 0\ 0\ 0\ 0\ 1\ 1\ 0\ 0\ 1\ 0$$

that Alice wants to send to Bob. Alice, before sending the message, encodes the message with a random sequence of bits, called the key, and then sends the encoded message through the communication channel, usually a fiber optics cable (Fig. 6.6). The random key is another sequence of bits known only to the sender (Alice) and the receiver (Bob):

$$\textit{Key}:\ 1\ 1\ 0\ 0\ 0\ 1\ 1\ 1\ 1\ 1\ 0\ 0\ 1\ 1\ 1\ 0\ 0\ 0\ 0\ 1\ 1\ 0\ 1$$

The transmitted sequence is obtained by adding the two sequences using the rules of binary addition, namely, $0 + 0 = 0, 0 + 1 = 1 + 0 =$

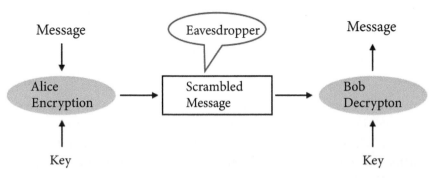

Fig. 6.6 Schematics for secure communication. Alice encrypts the message consisting of an array of 0's and 1's with a key. The encrypted message is sent to Bob who decodes it using the same key.

1, and 1 + 1 = 0. Using these addition rules, the transmitted sequence is thus a scrambled sequence of 1's and 0's:

Scrambled message: 0 0 0 1 0 1 1 0 0 0 1 0 0 1 1 0 0 1 1 1 1 1 1 1

The scrambled message is sent by Alice to Bob on a public channel. An important point is that the key is known only to Alice and Bob.

Once Bob receives the scrambled message encoded with the secret key, he unscrambles the message by adding the sequence representing the key, i.e.,

Key: 1 1 0 0 0 1 1 1 1 1 0 0 1 1 1 0 0 0 0 1 1 0 1

thus recovering the original message

Message: 1 1 0 1 0 0 0 1 1 1 1 0 1 0 0 0 0 1 1 0 0 1 0

The randomness of the key ensures that the transmitted data is also random and is inaccessible to a potential eavesdropper who does not have the key. The safety of the channel therefore depends critically on the secrecy of the key.

What we would therefore like in the present-day extensive electronic communication is to exchange a key (a random sequence of 0's and 1's) on a public channel (which is accessible to everyone including the resourceful eavesdropper) AND to be able to find out whether an eavesdropper, who can be located hundreds of miles away at an unknown location, has attempted to decipher the key. A protocol for quantum cryptography that achieves this objective was first proposed by Charles Bennett and Gilles Brassard in 1984. Consequently, this protocol is referred to as Bennett-Brassard-84 or simply BB-84. The Chinese satellite experiment was an implementation of this protocol.

6.4 Bennett-Brassad-84 (BB-84) protocol

We discussed earlier that the measurement of a quantum system in general causes a disturbance. In quantum cryptography, this aspect of quantum mechanics is used to allow two parties, Alice and Bob, to communicate in absolute secrecy, even in the presence of an eavesdropper, Eve.

There are two channels of communication in the BB-84 protocol: one is the quantum transmission channel through which an array of polarized photons is sent from Alice to Bob and the other is a classical channel like a telephone line or internet where Alice and Bob can exchange information about the preparation and measurement of the photons. It is essential that the eavesdropper, Eve, should not be able to block the classical channel and impersonate Alice to Bob and Bob to Alice. It is, however, assumed that Eve has unlimited resources to manipulate the photon in the transmission or quantum channel.

In the first step, Alice sends a beam of photons with each photon's polarization orientation being 0°, 45°, 90°, or 135° with the horizontal direction corresponding to states →, ↗, ↑, or ↖, respectively. The 0° and 90° orientations correspond to a ⊕ beam splitter whereas 45° and 135° orientations correspond to a ⊗ beam splitter. The two polarization states, say along 0° and 45°, stand for the bit "0" while the other two, along 90°, and 135°, stand for the bit "1." Alice encodes her sequence of data bits, switching randomly between ⊕ and ⊗ beam splitters. She then transmits the photons to Bob with regular time intervals between them.

Bob receives the photons and records the results using a random choice of ⊕ and ⊗ detection beam splitters. When the beam splitter orientations, chosen by Bob, is the same as that of Alice, the polarization of the received photon is perfectly correlated with Alice's photon. Thus, a photon polarized along 90° received through a ⊕ beam splitter will be found polarized along 90° and so on. However, no such correlation exists when the basis chosen by the receiver is different from that of the sender, i.e., a photon polarized along 90° will be found polarized either along 45° or 135° with equal probability if received through the ⊗ beam splitter.

As an example, let Alice send a stream of 9 photons with the sequence of polarizations along

$$0°, 90°, 135°, 0°, 45°, 135°, 45°, 45°, 90°$$

This sequence is sent by Alice choosing the beam splitter orientations

$$\oplus, \oplus, \otimes, \oplus, \otimes, \otimes, \otimes, \otimes, \oplus$$

This stream of photons, when received by Bob through a sequence of beam splitter orientations

$$\otimes, \oplus, \oplus, \otimes, \otimes, \otimes, \oplus, \otimes, \oplus$$

yields the outcome sequence with polarizations along

$$(45° \text{ or } 135°), 90°, (0° \text{ or } 90°), (45° \text{ or } 135°), 45°, 135°,$$
$$(0° \text{ or } 90°), 45°, 90°.$$

Bob records the outcome of his measurements in secrecy. Here notice that Bob's outcome is identical to that of Alice if he measures the photon in the same beam splitter orientation, $\oplus$ or $\otimes$, as that of Alice. However, when the beam splitter orientation chosen by Bob is opposite to that of Alice, then there is an equal probability for the two possible outcomes.

Next, Alice and Bob compare the orientations of their beam splitters for each photon through a conventional channel (such as a telephone line) without revealing the results. For the first photon, Alice tells Bob that her orientation was $\oplus$ and Bob tells Alice that his orientation was $\otimes$. Since the beam splitter orientation was different for Alice and Bob, they decide to discard the results for the first photon. For the second photon, the beam splitter orientations were identical for both of them, namely $\oplus$. Therefore, they decide to keep the outcome for the second photon, 90°. They do a similar comparison for the rest of photons and retain only the results where they use the same orientation of the beam splitters and discard the rest. Thus, in the above example, outcomes at 2, 5, 6, 8, and 9 are retained and the outcomes at 1, 3, 4, and 7 are discarded. When translated into bits 0 or 1 (1 0 1 0 1 in the above example), the key is obtained.

Thus, the first part is done: a key has been generated on a public channel with the two parties thousands of miles apart. However, before using this key, Alice and Bob would like to confirm that there was no eavesdropper seeking access to the key.

Next, we study what happens if there is an eavesdropper. The choice of Alice's and Bob's beam splitter orientations are completely hidden from Eve. A passive eavesdropping in this protocol is not possible as any attempt at eavesdropping would lead to discrepancies between the sequences.

As an example, consider the case when Alice sent a photon by choosing the orientation ⊕ with the polarization of the photon oriented along 90°. Also suppose that Bob received the photon in the same basis and his outcome for the polarization is along 90°. It is important to note that the orientation ⊕ is selected by Alice and Bob before they compared their sequence of orientations through a public channel which may be available to a potential eavesdropper. The eavesdropper located between Alice and Bob can intercept the photon and make any measurement on it when it passes through her. She has no way of knowing in what orientation Alice chose to send her photon and in what orientation Bob chose to detect the photon when it arrives at his end. Therefore, Eve has no way of knowing what basis Alice and Bob chose.

Eve thus has two choices when trying to infer about what is being transmitted. She can either use the same orientation, ⊕, that Alice and Bob used or the wrong orientation, ⊗. We now discuss these cases:

In the case Eve chooses the orientation, ⊕, and this may happen about 50 percent of the time, she gets the same outcome, 90°, that Alice chose and which Bob also detects.

However, in those roughly 50 percent instances when Eve chose the wrong orientation ⊗, she may get 45° or 135° orientations of the polarization with equal probability. If she gets a 45° orientation of the photon that she resends to Bob, the outcome at Bob's end will no longer be 90°, but, as a consequence of the discussion in the last section, it will be 0° or 90° with equal probability. The same happens if Eve finds the photon polarization orientation along 135°.

Thus, when Eve tries to eavesdrop with random orientation of her basis, roughly 25 percent of the outcomes at Bob's end will be different from those of Alice. Thus, Alice and Bob can try to infer the presence of an eavesdropper by comparing part of their data. If about 25 percent discrepancies are found, they can then be confident about the presence of an Eve who is trying to eavesdrop. They can then reject their data and start over.

Thus, the impossible has been achieved: exchanging a key on a public channel in such a way that an eavesdropper can be traced with almost 100 percent accuracy. This is, thus, a fascinating application of quantum mechanics.

6.5 Quantum money

An interesting and potentially important application of quantum security is quantum money. This idea, due to Stephen Wiesner, was first presented in 1983, and was a precursor to the BB-84 protocol for communication security discussed above. The problem is how to design a fully secure system to avoid counterfeiting of currency notes. More specifically, if we have a currency note, a counterfeiter could try to make an identical copy. Is it possible that, regardless of how good a forgery, the copy is identified as such with certainty?

In the present world, tremendous efforts have been made to make the task of the counterfeiter very difficult. Each currency note has embedded strips, holograms, special inks, or microprinting to safeguard its integrity. However, from a classical perspective, no matter how sophisticated our techniques, it is impossible to absolutely guarantee that a counterfeiter cannot succeed for the simple reason that any printing device a good guy can build, a determined bad guy can also build an identical copy.

The quantum aspects of the photon, as discussed above, can however make it possible to potentially create a currency note that cannot be copied and it should be possible to identify a forged copy with certainty.

Typically, a currency note is identified only by a serial number. But, in a quantum currency, each bill can have, in addition to a classical serial number, n photons trapped secretly prepared in one of the four states, $\rightarrow, \uparrow, \nearrow, \nwarrow$. At the bank, there is a record of all the polarizations and the corresponding serial number as shown in Fig. 6.7. On the bank note, the serial number is printed, but the polarizations are kept secret. The

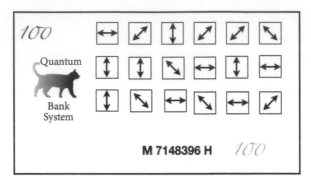

Fig. 6.7 A quantum currency note has a serial number and a number of trapped photons each stored in one of the four states →, ↑, ↗, ↖.

bank can always verify the polarizations by measuring the polarization of each photon with the correct orientation of the beam splitters without introducing any disturbance.

For a potential forger, the serial number is known, but the challenge is how to copy the polarization states. He needs to know the orientation, $\oplus$ and $\otimes$, to correctly measure the state of the photon and then copy these photon polarization states on his forged currency note. He faces the same dilemma that Eve faced in the quantum cryptography protocol. If the forger uses the correct orientation of the polarization beam splitter (and this may happen 50 percent of the time), he can find the secret polarization state and copy it faithfully on his forged copy. However, if he chooses the wrong orientation (and this happens also about 50 percent of the time), it will change the polarization of the photon in the trap. The forged banknote created will be with this wrong polarization.

Let us illustrate this with an example. We assume that the polarization state of the photon in the currency note is →. If the forger measures it with orientation $\oplus$, he gets the correct polarization state → and his forgery may not be detected. However, if he measures it with the orientation $\otimes$, he gets either the state ↗ or the state ↖ with equal probability. Let us assume his outcome is ↗. Now at the bank, when polarization is checked with the correct orientation $\oplus$, there is a 50 percent chance that the outcome will be →, but also a 50 percent chance that the outcome will be ↑. Thus, for each photon the success probability is 3/4 in duplicating it correctly but a failure probability of 1/4. If the number of stored photons is large, this probability can become almost equal to zero.

A forgery can therefore be avoided in principle using the basic ideas of quantum mechanics. A practical implementation of this idea is, however, difficult mainly because of the inability to store definite states of polarization of photons for a long time. Still, the idea demonstrates at least the potential of using a counterintuitive aspect of quantum mechanics for something very practical in a forceful manner.

7

Does God play dice?

*Quantum mechanics is certainly imposing. But an inner voice tells
me that it is not yet the real thing. The theory says a lot but does
not really bring us closer to the secret of the "Old One." I, at any
rate, am convinced that He is not playing at dice.*

— *Albert Einstein*

Einstein, stop telling God what to do.

— *Niels Bohr*

Perhaps the most celebrated and discussed aspect of quantum mechanics
is the probabilistic nature of its predictions. As we have seen, the particle
trajectory is a central concept in Newtonian physics. For example, if a
ball is thrown toward a wall, its position and how fast it will be moving
at a later time can be determined with arbitrary accuracy using Newton's laws of motion. The precise position at which the ball will hit the
wall can also be predicted. The motion of the particle is therefore fully
deterministic and follows a known trajectory.

An amazing result is that such a deterministic behavior is no longer
valid according to the laws of quantum mechanics. We cannot predict
with absolute certainty where the ball will be and how fast it will be moving after a given time and we cannot give an absolutely precise location
where the ball will hit the wall. We can only give a probabilistic answer
like "There is a 99 percent probability that the ball will hit the wall at a
certain height." This probability can be very close to 100 percent (such
as 99.9999 percent) but never equal to 100 percent. This seems very surprising as this is contrary to our common observations which appear to
be explained in a fully satisfactory manner by Newtonian mechanics.

Before discussing the motion of a large object like a ball or a small object like an atom or electron, we consider a simple example like the passage of a photon polarized in a certain direction passing through a polarization beam splitter. This example should illustrate the probabilistic nature of quantum mechanical laws and their consequences.

7.1 Two-state examples

In the last chapter, a simple two-state system was described with amazing properties. The polarization property of a single photon exhibits a behavior that we do not expect on the basis of our classical intuition. For example, a photon polarized in the vertical direction (↑) passes through a polarization beam splitter with polarization in the vertical direction. However, the same photon, when it passes through the polarization beam splitter rotated by 45 degrees, has a 50 percent probability each to come out with polarization oriented at 45 degrees (↗) and at 135 degrees (↖). The laws of quantum mechanics can only give the probability (50 percent each) for passage with either orientation (Fig. 7.1).

This inability to make a definite prediction about the polarization direction has been a matter of debate for almost a century and scientists are still trying to understand it. The probabilistic nature of quantum mechanical predictions led Albert Einstein to declare that quantum mechanics is an "incomplete" theory. In his viewpoint, a "complete"

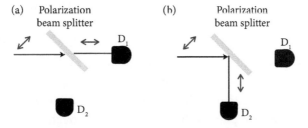

Fig. 7.1 Two photons in identical state ↗ pass through the polarization beam splitter. (a) One photon is detected at detector D_1 in the horizontally polarized state → and (b) the other at detector D_2 in the vertically polarized state ↑.

theory should be able to give definite predictions. Newtonian mechanics does that but quantum mechanics can only tell us, as in the above example, whether an event will happen or it will not probabilistically. Einstein therefore considered quantum mechanics to be incomplete.

7.2 Hidden variables

In order to illustrate this point further, we consider, instead of a single photon, two photons that are identical in every respect and prepared in the vertically polarized state ↑. Let these photons pass one by one through the polarization beam splitter rotated by 45 degrees. According to the discussion above, the first photon has two possibilities with 50 percent probability each: either pass through with a polarization along 45 degrees, in state ↗, or with a polarization along 135 degrees, in state ↖. The same is true for the second photon.

We thus have four possible outcomes with equal probabilities for the two photons: both are found in the state ↖, the first is found in the state ↖ and the second is found in the state ↗, the first is found in the state ↗ and the second is found in the state ↖, and both are found in the state ↗.

The situation is similar to flipping two coins where we have 25 percent probability each that we get both coins with head up, the first with head up and the second with tail up, the first with tail up and the second with head up, and both with tail up. However, in this case, if we knew all the forces and the initial orientations of the coins, we could predict with certainty whether we will get head up or tail up. The reason why the outcome appears random is due to our ignorance of the initial conditions. This is not the case with quantum objects like the polarization state of a single photon. There is no way that we can distinguish the two photons before they pass through the beam splitter, yet it is impossible to predict in advance which polarization direction it will take after passing through it.

Let us suppose that the first photon is found in the state ↖ and the other in the state ↗. This can happen with 25 percent probability. First, we ask the obvious question: What is the difference between the two photons? The equally obvious answer is that the first photon after passing through the polarizing beam splitter is found in state ↗ and the other in state ↖. But then we ask the difficult question: What was the difference between the two photons before they passed through the beam splitter? If, somehow, we knew the difference, we could understand why the two photons behaved differently. However, as discussed above, according to quantum mechanics, there was absolutely no difference between the two photons, yet one of them ended up in state ↖ and the other in state ↗.

This is very mysterious. This is the probabilistic aspect of quantum mechanics that Einstein never accepted, proclaiming that "God does not play dice." In view of the inability of quantum mechanics to give a definite answer, Einstein declared quantum mechanics to be incomplete.

One way out of this conundrum is to assume that the two photons, being apparently identical in all respects, were indeed different even before they entered the beam splitter. Einstein conjectured in 1935 that these photons had some "hidden" properties that we do not know and we cannot measure in the laboratory but which distinguished them from each other. This is akin to having a ghost in the laboratory: if the ghost, that we can neither see nor feel directly or indirectly, points to the left, the photon will be polarized along 45 degrees and if, it points to the right, photon will be polarized along 135 degrees. If some day, we are able to identify these "hidden variables" and incorporate them into the theory, then we will be able to make definite predictions like in Newtonian mechanics. Quantum mechanics would then become "complete."

The ultimate answer about the completeness or incompleteness came 30 years later, almost ten years after Einstein's death, and the answer was nothing that Einstein would have expected. More about it in Chapter 12 when we discuss the questions of *reality*.

7.3 Schrödinger equation

After this simple example that dwelt on the probabilistic nature of quantum mechanics, we turn to the question of how the concept of trajectory and determinism disappeared with the advent of quantum mechanics.

Perhaps the most defining moment in our understanding of the laws of nature came in January 1926 when Erwin Schrödinger formulated quantum mechanics and wrote the equation that came to be known as the Schrödinger equation. The story how Schrödinger got to change our perception of the cosmos through this epoch-making equation has been related by Felix Bloch, the first Ph.D. student of Werner Heisenberg and the winner of the 1952 Nobel Prize. He was a student at the Swiss Federal Institute of Technology (ETH) in Zurich in 1925. In his reminiscences on the occasion of the 50th anniversary of quantum mechanics, he described the birth of the Schrödinger equation in these words (*Physics Today*, December 1976):

> Once at the end of a colloquium I heard Debye saying something like: "Schrödinger, you are not working right now on very important problems anyway. Why don't you tell us some time about that thesis of de Broglie, which seems to have attracted some attention." So, in one of the next colloquia, Schrödinger gave a beautifully clear account of how de Broglie associated a wave with a particle and how he could obtain the quantization rules of Niels Bohr ... by demanding that an integer number of waves should be fitted along a stationary orbit. When he had finished, Debye casually remarked that he thought this way of talking was rather childish ... to deal properly with waves, one had to have a wave equation. It sounded quite trivial and did not seem to make a great impression, but Schrödinger evidently thought a bit more about the idea afterwards. Just a few weeks later he gave another talk in the colloquium which he started by saying: "My colleague Debye suggested that one should have a wave equation; well, I have found one!"

The equation Schrödinger found is one of the most important equations of science of all time. Its importance and impact cannot be overstated. Schrödinger's equation replaced Newton's equations of motion as the fundamental law of nature.

Whereas Newton's laws gave birth to the determinism that dominated the science of the eighteenth and nineteenth centuries, Schrödinger's equation spelled an end to determinism and heralded the birth of the era of probabilistic universe. This work would eventually lead to consequences that were so strange and counterintuitive that not only made giants like Einstein very uncomfortable but, by some accounts, made Schrödinger himself regret being the founding father of quantum mechanics. More about it in Chapter 11 on Schrödinger's cat and entanglement.

According to quantum mechanics, a "particle" (e.g., a tennis ball or an electron) does not follow a definite trajectory. The complete description of a particle is contained in a quantity called the "wavefunction," typically represented by the Greek letter psi (pronounced as "sy"), ψ. This quantity describes the spatial and temporal behavior of the "particle." In Newtonian mechanics, the object like a ball is characterized by its location and the speed with which it is moving. Quantum mechanically, the same object is described by the wavefunction. The wavefunction ψ is a mathematical object that depends on the position, speed, and other attributes of the object. The wavefunction contains all the information about the object. If we know ψ, we can determine any observable property (e.g., location, speed, ...) of the object. The importance of the Schrödinger equation is that it provides the tools to determine ψ computationally and then to use ψ to determine the properties of the system.

7.4 End of determinism

Immediately after Schrödinger derived the equation for ψ in 1926 and correctly derived the properties of the simplest known physical system, the hydrogen atom, the debate began about the meaning of the wavefunction. How do we interpret the wavefunction ψ? This was one of the most important questions facing the physicists immediately after Schrödinger formulated his equation for the wavefunction.

The breakthrough came through the work of Max Born. In June 1926, he presented the probabilistic interpretation of ψ according to which the square of the wavefunction, $|\psi|^2$, at any point in space, is proportional to the probability of finding the particle at that point. The wavefunction, ψ, itself has however no physical meaning. This realization due to Max Born amounted to the end of the era of certainty that we associated with the predictions in Newtonian mechanics.

So, whereas, on the basis of Newtonian mechanics, we can predict with certainty where a tennis ball will land on a wall if we knew the initial location, speed, and the force of gravity, we have no such prediction for an electron that follows exclusively quantum mechanical laws. Quantum mechanics does not allow us to describe the motion of an electron in terms of a trajectory and does not tell us where the electron will hit the screen. Quantum mechanics only gives us the probability that the electron will hit a certain point on the screen.

An important point to note is that the dynamics of the system based on the Schrödinger equation is deterministic just like it is according to the Newtonian equation of motion. However, there is one basic difference. In Newtonian mechanics, we can predict with certainty the values of the physical variables like the position and the speed of a particle at a later time if we knew these values in the present. However, the Schrödinger equation can provide the precise value of the wavefunction for all points in space at a later time if we knew the value of the wavefunction at all those points at the present time. The probabilistic nature of quantum mechanics is embedded in the fact that the wavefunction does not give us a definite value of position and speed of the particle, it only gives the probability of finding the location and the speed of the particle. Thus, if the value of the wavefunction is given over an area and we make a measurement, the particle will be found randomly at any point on the surface. There is no prediction for a single measurement. However, if make these measurements over a large number of identically prepared systems, the particle will be found more times at locations where the value of the square of the wavefunction is large.

An amazing outcome is that, not only did the Schrödinger equation explain all the scientific facts known at that time, but it could predict startling new effects that could be tested in the laboratory. A new revolution in science had begun whose consequences shaped life in the twentieth century and continues to do so today.

The dynamical equations for a particle in quantum mechanics and classical mechanics are very different and there is no resemblance between the Schrödinger equation and the equation of motion according to Newton's laws of motion. So how do we understand the connection between the two? We do not see the probabilistic nature, that is central to quantum mechanics, in our everyday life. Newtonian mechanics gives very accurate and precise results. How do we reconcile these observations?

In order to answer these questions, we consider the simplest problem of particle dynamics and study it in both Newtonian mechanics and quantum mechanics and see where the fundamental differences come in the two approaches. This simple problem also helps to illustrate that the fundamental theory is quantum mechanics and Newtonian mechanics is an approximation, a remarkably good approximation, when considering big objects.

7.5 Schrödinger vs Newton

The problem we consider is that of an object at some initial location and moving with a constant speed in a given direction. The question posed is: What is the location of the object at a later time?

The answer to this question in Newtonian mechanics is simple and straightforward. The object will follow a simple trajectory and we can predict where it will be after one second or two seconds later very precisely. This is the hallmark of Newtonian mechanics. Since the external force is equal to zero in our example, the object will continue to move with the same speed. If a ball is moving at a uniform speed of 1 meter per second in the horizontal direction, it will travel 1 meter in one second, two meters in two seconds, three meters in three seconds, and so on. This is shown by the dots in Fig. 7.2.

What about the solution of the same problem within the framework of quantum mechanics? The corresponding equation for the wavefunction ψ for a free particle is the Schrödinger equation. The square of the wavefunction gives the probability of finding the object at a given location at a given time. This is quite a sophisticated problem that requires knowledge of advanced mathematical techniques. What we find is that the object, instead of following a well-defined trajectory, is given by a probability distribution that keeps spreading with time. Where is the object located at a later time? We cannot give a deterministic answer to this question. We can only give the probability of finding the object at any point in space at a given time. The most probable location of the particle is the same as predicted by Newton's law of motion but it can also be found far from the Newtonian trajectory, though with very small probability. As

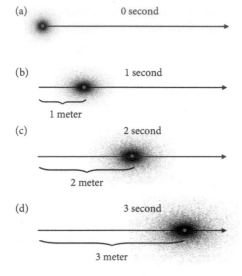

(a)   0 second

(b)   1 second

1 meter

(c)   2 second

2 meter

(d)   3 second

3 meter

Fig. 7.2 The motion of an object when no force acts on it. The predictions of Newtonian equations of motion are the dots as the time goes by. (b) The fuzzy regions, that become broader as the time elapses, represent the probability distribution, $|\psi|^2$, of finding the object according to the quantum mechanical laws.

the particle moves, the corresponding probability distribution spreads and the region where it can be found increases as shown in Fig. 7.2. This is a startling result that our commonsense is unable to believe. In principle, if a tennis ball is moving with no force acting on it, there is a probability, albeit extremely small, that it can be found even one mile away.

But then how come we always find a well-defined trajectory for any moving object and no deviation is observed from the path that Newtonian mechanics predicts? There are two reasons. The first is that the probabilistic description is valid only if an object is not looked at, whereas a moving object, such as a tennis ball, is constantly being looked at. The act of observation continuously collapses the wavefunction to a localized point, thus yielding a trajectory. In the words of Werner Heisenberg:

> I believe that the existence of the classical "path" can be pregnantly formulated as follows: The "path" comes into existence only when we observe it.

Secondly, even if the object is not looked at, the spreading of the probability distribution is extremely slow even for as tiny an object as a dust particle. We shall have to wait for a long time, longer than the age of the universe, for a perceptible spreading of the probability distribution for such a tiny object. The is due to the extremely small value of Planck's constant as discussed in Section 3.7. Such spreading can, however, be observed readily for objects of the size of an atom or electron.

The amazing picture that emerges for a simple motion of any object such as a tennis ball is as follows: At any time, the ball cannot be described as an object which is localized at a well-defined position if it is not being looked at. Instead, it is like a cloud of something which is described by what we call the wavefunction. This cloud is spread over a large region but it is not something real. This cloud is dense in the middle but gets thinner as the distance from the center increases. As the ball moves, the density of the cloud gradually becomes more spread out in the direction of motion. This evolution continues as long as the cloud is not looked at or measured. As soon as the tennis ball is looked at, the cloud that is spread over a large region *collapses* to a single location instantly. The probability where it collapses is proportional to the density of the cloud at that location.

This incomprehensible picture has many fundamental questions embedded in it. Where was the tennis ball located *before* we looked at it? Was there even a real ball before the measurement or the observation took place? At what moment did the *collapse* take place? What about causality? Just before we looked at the ball, the probability of existence of the ball at a large distance away (in principle a mile away) was non-zero. But when looked at it, this probability collapsed at a single point instantaneously.

Quantum mechanics deals with observable quantities. The observable properties like position, speed, polarization have meaning only when they are observed and measured. But what about the quantities that cannot be measured? Do they not have any meaning or reality? For example, do concepts like the trajectory of a ball or the orbits of electrons in an atom not exist until we observe them?

These and other questions have sparked a debate about our understanding of nature that continues to this day with no definite answers in sight. However, the probabilistic predictions of the quantum mechanical equations have an impeccable record—there is not a single prediction of quantum mechanics that has been proven wrong.

The demise of a certain and deterministic world with the advent of quantum mechanics is a major intellectual leap in our perception of this universe.

8

An uncertain world!

The uncertainty principle "protects" quantum mechanics. Heisenberg recognized that if it were possible to measure the momentum and the position simultaneously with a greater accuracy, the quantum mechanics would collapse. So he proposed that it must be impossible. Then people sat down and tried to figure out ways of doing it, and nobody could figure out a way to measure the position and the momentum of anything—a screen, an electron, a billiard ball, anything—with any greater accuracy. Quantum mechanics maintains its perilous but still correct existence.

— Richard Feynman

Louis de Broglie's description of particles, such as an electron, as a wave is very intriguing, to say the least. However, this description is validated by the electron diffraction experiment of Clinton Davisson and Lester Germer that was done immediately after de Broglie presented his hypothesis in 1924. Later, in Chapter 10, we discuss another landmark experiment that shows electrons exhibiting interference in a double-slit experiment. But then what about the innumerable experiments where electrons exhibit their particle nature? For example, we use the classical laws of physics to describe the electron deflection in the presence of electric and magnetic fields. There, we treat an electron like a particle with definite mass and charge.

A natural question is: How do we reconcile both wave and particle natures of an electron, as well as any other massive particle? A related question is: How do we describe a particle at a particular location either at rest or moving with a certain speed within the context of wave–particle duality? These questions lie at the heart of quantum mechanics and the answer to these questions are rigorously given by the wavefunction approach. Here we motivate such an approach via describing an electron as a packet of waves. This approach is not only helpful in understanding the wavefunction approach of quantum mechanics but also useful in understanding Heisenberg's uncertainty relation and Bohr's principle of complementarity that we discuss later in this chapter.

It was known for a long time, well before the advent of quantum mechanics, that a time signal can be decomposed into a distribution of frequencies. For example, a piece of music is composed of different frequencies generated by musical instruments. The frequency of the wave generated by a drum is smaller and by a flute is larger. A suitable combination of these frequencies can generate beautiful music. A piece of music, indeed any sound signal, can either be analyzed as a signal varying in time or equivalently as a combination of waves of different frequencies. In the same way, a specific sound can be looked at two different ways, either as a time signal or a combination of different frequencies. It is also well-known that a sharp time signal, such as a beep, is composed of a large number of frequencies, whereas a smooth sound spread over a long time period is composed of only a few frequencies.

Similar ideas apply to a localized object. An object can be looked upon as occupying a finite space. Another way of looking at it is by assuming that the object is composed of what we call spatial frequencies (or wavelengths). A spatial frequency is similar to a sound frequency, but in space. For example, if there is a long row of marbles, with a distance of two inches between the two neighboring marbles, the spatial frequency will be ½ marbles per inch. A single localized marble can be described by adding an infinite number of such spatial frequencies. This is the basic idea behind the description of a particle as being composed of a large number of spatial waves of different frequencies or wavelengths. An object is therefore described in terms of a wavefunction consisting of many spatial waves.

The bizarre wavefunction description of an object is nothing that we encounter in our everyday life, yet this is the underlying picture of everything that exists in nature. At the most fundamental level, everything in the universe is described by a wavefunction.

8.1 What does a wavefunction look like?

As discussed earlier, a wave is described by two parameters: the frequency which is the number of crests that pass a point per second and the wavelength which is the distance between two neighboring crests. An important property of the wave is the principle of superposition: if there are two or more waves at a location at a given time then the waves add at each point.

Armed with this description of waves, we can see how waves can be combined together to get a localized object like an electron.

According to the de Broglie hypothesis, a particle can be described by a wave, the de Broglie wave. However, a single wave extends to a very large region of space, in principle, over an infinite region. But a particle is confined in a very small region. A small particle, like a dust particle or even an electron, is localized in one spot. How do we reconcile these two completely opposite descriptions for the same object?

We can pose the question somewhat differently. We have seen that a particle is described by a wavefunction. What does the wavefunction look like for a particle? The answer is that a wavefunction consists of a large number of de Broglie waves of varying wavelengths. A localized particle is therefore described by a superposition of de Broglie waves with multiple wavelengths.

As the simplest such example, we consider two waves of equal amplitudes but with slightly different wavelengths. The superposition of the two waves with close wavelengths together breaks up the continuous wave into a series of packets (Fig. 8.1). As the number of waves increase, the superposition can lead to constructive interference in a small region as shown in Fig. 8.2.

To describe a single electron (or any particle) confined to a highly localized region, we need a single wave packet that is zero or nearly zero everywhere in space except for one localized region. Such a wave packet can be constructed by superposing waves having a continuous distribution of wavelengths, centered around a certain wavelength. In this case, the waves become out of phase after a short distance and interfere destructively. The result is that a single wave packet is obtained.

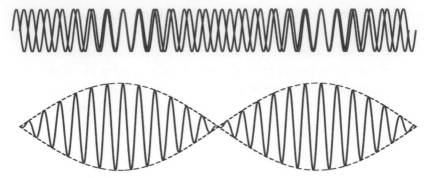

Fig. 8.1 Superposition of two waves with slightly different wavelength. The regions where the two waves have the same phase, a constructive interference leads to a large amplitude and where they are out of phase, destructive interference leads to a zero amplitude.

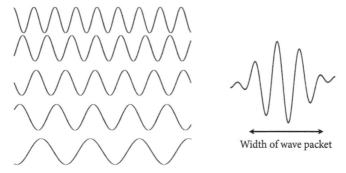

Width of wave packet

Fig. 8.2 Superposition of 5 waves leads to the formation of a wave packet. The width of the wave packet depends on the wavelength of the waves involved. A wave packet consisting of waves of smaller wavelengths will be narrower.

In Fig. 8.3, we show a superposition of 3, 5, 9, and an infinite number of waves, centered around a wavelength. The wave packet shown in 8.3(d) is clearly particle-like in that its region of significant magnitude is confined to a localized region in space. The width of the wave packet depends on the wavelength of the waves—the smaller the wavelengths involved, the narrower and well localized is the wave packet.

The wave packet provides the information about the probability of finding the particle at a certain position. Thus, the probability is larger

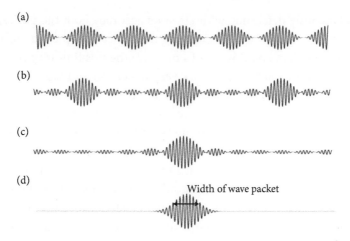

Fig. 8.3 Superposition of (a) 3, (b) 5, (c) 9, and (d) infinite number of waves.

where the amplitude is larger and is zero where the amplitude is equal to zero. Here we recall that the probability is proportional to the square of the wavefunction.

In this picture, we are able to answer the question as to why we see the particles localized at a very well-defined position in our everyday life. Just imagine that the waves that form the wave packet are of extremely short wavelengths. Then the width of the wave packet becomes extremely small. The wavefunction is non-zero in a very narrow region and almost zero elsewhere. This corresponds to a very well localized object.

8.2 Heisenberg uncertainty relation

An important consequence of quantum mechanics is the Heisenberg uncertainty relation. According to these relations, there are complementary pair of variables such that if we know one of them very precisely then the other becomes uncertain.

A pair of complementary observables is the position and the speed of a particle at a given time. According to Heisenberg, it is impossible to simultaneously determine with absolute accuracy both the position of the particle and how fast it is moving no matter how precise our measurement devices are. If we try to measure the position very precisely, the speed of the particle becomes very uncertain and vice versa. This is again in sharp contrast to Newtonian mechanics where we can measure any observable property as precisely as we like; the only restriction comes from the limitation of our measurement devices.

Why do we not observe this behavior in everyday life? Usually, we are able to see an object located at a precise position and can determine how fast it is moving with almost arbitrary accuracy. The Heisenberg uncertainty relation does not appear to be valid. The reason is that the product of the uncertainty in the position (region in which the object can be found) and the uncertainty in its speed is equal to or greater than the value of Planck's constant. As learnt in Section 3.1, Planck's constant is incredibly small. Therefore, for big observable objects, the uncertainties in both position and the speed can be extremely small, smaller than the resolving power of most measuring instruments. Only for microscopic objects like an electron or an atom does the uncertainty relation lead to observable effects.

How do we understand the Heisenberg uncertainty relation from the wavefunction approach of quantum mechanics? We qualitatively discuss it from two points of view.

A simple way of understanding the uncertainty relation is via diffraction through a single slit. We recall that when a beam of light passes through a slit, the width of the beam no longer stays the same. Instead, the beam spreads. This is shown in Fig. 8.4. The spread is large for narrow slits and small for wide slits. For a very wide slit, there is almost no spreading. How do we understand the uncertainty relation from this simple observation?

Instead of a light beam, let us consider the passage of an electron through the single slit. As seen above, an electron can be described by a wavefunction consisting of de Broglie waves. Thus, electrons behave like light waves and are diffracted in the same manner as the light waves. We now like to see how best can we measure the position and the speed of an electron, say in the vertical direction, perpendicular to the direction of the propagation of the electrons. The best way to find the

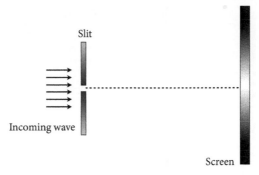

Fig. 8.4 Light is incident on a slit. After passing through the slit, the width of the beam increases due to diffraction. The intensity of light is maximum at the center but decreases gradually as we move away from the center.

position of the electron is to pass it through a narrow slit. The width of the slit determines the uncertainty in the location of the electron. The narrower it is the more precise information we have about the position of the electron when passing through the slit.

But what about its speed in the vertical direction? We first note that, if the electron is detected at the central point on the screen there is no deflection and the vertical component of the speed of the electron is zero. However, if it is detected away from the center, the vertical component of the speed must be non-zero. Let us now return to the question about the spread or the uncertainty in the speed of the electron in the vertical direction. For a very narrow slit, the wavefunction of the electron spreads much more just like the light passing through a very narrow slit. This means that there is a large range of region where the electron can be detected on the screen. This in turn means that the uncertainty in the speed in the vertical direction becomes large. Thus, we have the complementary behavior. If the slit is narrow, we can measure the position very precisely and the uncertainty in the position of the electron is low. However, in this case, the wavefunction of the electron spreads and consequently the speed component in the vertical direction becomes very uncertain. Similarly, if the slit is wide, we do not know where the electron passed and consequently the position of the electron is very uncertain. However, in this case the spread of the wavefunction is very small, and

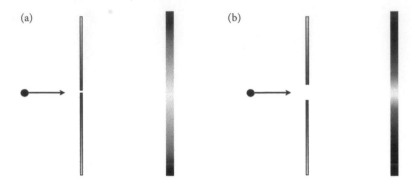

Fig. 8.5 An electron, after passing through the slit, hits the screen. In (a) the probability of finding the electron is plotted for a narrow slit and in (b) for a wider slit.

we can be quite certain how fast it must be moving. These situations are depicted in Fig. 8.5.

An alternative way of seeing the Heisenberg uncertainty relation is based on the wavefunction approach discussed in Section 8.1. According to the de Broglie description, the speed of the particle is directly related to the corresponding de Broglie wavelength. If the wavelength is small, the speed of the wave is large and when the wavelength is large, the speed is small. What we saw in the previous section is that a particle, like an electron, is described by a wavefunction. The full wavefunction is formed by combining a large number of de Broglie waves of varying wavelengths. A localized particle is therefore made up of a large number of de Broglie waves.

A very precise location means that the wave packet consists of a very large number of de Broglie waves of very short wavelengths. However, each wavelength corresponds to a different speed of the particle. There are, therefore, a large number of possible speeds of the particle leading to a large uncertainty in the speed of the particle. In the opposite case, if the particle has a well-defined speed, there should be a very small number of de Broglie waves, ideally a single de Broglie wave. However, in this case there is no localization—the particle can be anywhere.

Thus, we can conclude on the basis of this simple argument that if a particle has a precise location, its speed is highly uncertain and if it has a precise speed, its location is highly uncertain. This is the essence of the Heisenberg uncertainty relation. It is clear that any attempt to define both the position and the speed of a particle at the same time will be futile. A careful mathematical analysis shows that the product of the uncertainties in position and speed is equal to Planck's constant h divided by the mass of the particle.

An obvious question is why we do not see the consequences of the uncertainty relation in our everyday life? How come we are able to measure the position and speed of a particle, any particle, to an arbitrary accuracy simultaneously? The reason, as mentioned above, is that Planck's constant is extremely small and the uncertainty relation becomes relevant only for objects that are very small, like electrons and atoms. To see this, we recall that Planck's constant h is equal to 6.26×10^{-34} kilogram-meter-meter per second (Section 3.7). Even for a dust particle of mass 10^{-12} kilogram, the product of the uncertainties between position and the speed is about 10^{-21} meter-meter per second. This is an incredibly small number. We can measure the position of the dust particle to an accuracy to (say) one millionth of a meter and the uncertainty in the speed to one trillionth of meter per second. For a bigger object like a tennis ball, the uncertainties are even smaller. For big objects, we need extreme precision measuring devices to measure the position and speed. Such measuring devices are not presently available.

8.3 Measurement disturbs the system

Next, we address the question: What is the origin of the Heisenberg uncertainty relation? Here we only mention that the process of measurement is an inherent source of uncertainty. According to Heisenberg (in his 1927 paper):

> If one wants to be clear about what is meant by "position of an object," for example of an electron . . ., then one has to specify definite experiments by which the "position of an electron" can be measured; otherwise this term has no meaning at all.

If, for example, we want to measure the position of an electron we have to come up with a device like a microscope to precisely measure the location of the electron. This can be done by first shining light on the electron and measuring its position by looking at the scattered light. However, a consequence of wave–particle duality is that the light consists of photons which behave like particles. Just like any other particle, when hitting the electron, the photon disturbs both its location and its speed.

Thus, when photons hit the moving particle, the speed of the particle changes randomly. The consequence is that if we try to measure the position of a particle very precisely, its speed changes randomly. Similarly, we can argue that if we measure the speed of the electron precisely, its location becomes uncertain. Heisenberg could show a relationship between the preciseness of measurements of both position and momentum. Heisenberg's uncertainty principle, formulated in 1927, is one of the cornerstones of quantum mechanics. It is based on the principle that it is impossible to measure anything without disturbing it.

Let us consider an interesting manifestation of the uncertainty relation, not for microscopic particles like atoms and electrons, but for a big object like a ping pong ball. Let a ping pong ball drop on another ping pong ball dead on. We assume that everything is perfect—no wind is blowing, there is no friction between the two balls, the lower ball is at rest etc. The question is how many bounces the ping pong ball will make? Newtonian mechanics tells us that, under the ideal conditions, there will be an infinite number of bounces as there is no force except the force of gravity. What about the answer to the same question within the quantum mechanics framework? There will be an uncertainty for the ball in the position and how fast it is moving when it hits the ball on the ground. There is, therefore, a probability that the ball will not hit head on but slightly on the side. This effect will be magnified with any subsequent bounces. The result is that, after certain number of bounces, the ball will hit so far away from the center that it will not be able to bounce back again. A careful calculation shows that, remarkably, the maximum number of bounces is nine.

8.4 Bohr's principle of complementarity

When Heisenberg formulated his uncertainty relations, Niels Bohr formulated the principle of complementarity. According to this principle, two observables are *complementary* if precise knowledge of one of them implies that all possible outcomes of measuring the other one is equally probable.

We have seen above that light and electrons can behave both as particle or wave in different experiments. For example, light can behave like a wave in interference and diffraction experiments but behaves like a particle (quantum of energy) in experiments such as photoelectric emission. According to the principle of complementarity, light (and electrons) can behave either as a wave or a particle but never both in a given experiment.

Such behavior is exhibited by photon polarization. When a vertically polarized photon passes through a polarizing beam splitter oriented by 45 degrees, the outcome is equally probable for the polarization orientations 45 degrees and 135 degrees. Similarly, if a photon polarized along 45 degrees passes through the polarizing beam splitter, it has equal probability of passing through in a horizontally or vertically polarized state. We can combine these statements as follows: if the polarization of the photon is known in a particular direction, then all the possible outcomes of measurement via a polarizing beam splitter rotated at 45 degrees with respect to the photon polarization are equally probable. This is a statement of Bohr's principle of complementarity for a two-state system.

The basic idea of complementarity is that there are two ways to look at reality and both can be true. However, they cannot both be seen at the same time. The uncertainty relation and complementarity appear to be related concepts. However, they have been shown to be quite different from each other.

An indirect consequence of the principle of complementary and the Heisenberg uncertainty relation is that an unknown quantum system cannot be copied or cloned.

8.5 Can we clone quantum systems?

Heisenberg's uncertainty relation and Bohr's principle of complementarity form the foundational principles of quantum mechanics. If these are violated then the edifice of quantum mechanics can come crashing down. Thus, any process that can potentially lead to a violation of these *sacred* principles must be examined with utmost care. One such process is the cloning of the quantum states.

It is our common experience that, given the expertise and resources, we can make identical copies of any object. In general, an object can be copied such that the copy is indistinguishable from the original. For example, a document can be copied such that the original and the copy are identical. But is this true even at the microscopic level? Can we build a cloning machine that can make an identical copy of a single photon or an electron? The answer is, surprisingly, No.

The question of interest is whether it is possible to make a perfect copy or a clone of an unknown quantum state without destroying the original state. If this becomes possible then we could make as many copies of the quantum state as we like. We can then make measurement of any variable with arbitrary precision, leading to a violation of both Heisenberg's uncertainty relation and the principle of complementarity.

In order to illustrate this point, let us go back to the example of the measurement of the polarization of photons in different bases as discussed in Chapter 6. We noted that a photon, when observed in the $\oplus$ basis can either be in the state ↑ or in the state →. However, the same photon, when observed in the $\otimes$ basis can either be in the state ↗ or in the state ↖. The principle of complementarity does not allow us to measure the polarization of the photon in the two bases simultaneously. However, if cloning becomes possible, we can make identical copies of the photon. We can then measure the polarization of half of them in the $\oplus$ basis and the other half in the $\otimes$ basis. These measurements can give precise values of the complementary variables, thus violating the principle of complementarity.

In a similar way, we can show that, if cloning of quantum states is allowed, Heisenberg's uncertainty relations can be violated as well.

The no-cloning theorem, showing that cloning of an unknown quantum state is not allowed, was formulated in a classic paper by William Wootters and Wojciech Zurek in 1982. The foundation of quantum mechanics is therefore protected. The motivation for this paper came when a paper by Nick Herbert was published in 1982 showing that photon cloning can lead to superluminal (faster than light) communication, thus violating another tenet of modern physics, namely, Einstein's theory of relativity. According to the theory of relativity, no information can be sent faster than the speed of light.

9

Vacuum is not "nothing"

When you look at a vacuum in a quantum theory of fields, it isn't exactly nothing.

— *Peter Higgs*

Before the advent of quantum mechanics, vacuum was perceived as *nothing*—a place where no light existed, nothing moved, and there was no energy present. One of the most dramatic predictions of quantum theory is that vacuum has an infinite amount of energy and new "virtual" particles are constantly being created and destroyed.

As we discussed earlier, light has a dual character. In some experiments it can act like a wave and in some others like particles. A synthesis of these two dramatically different behaviors came through the work of Paul Dirac.

As mentioned before, light consists of different colors, visible and invisible. All these colors are distinguished by the frequency which is defined as the number of the crests passing through a point each second. These frequencies can take on any value and are, therefore, infinite in number. In the quantum mechanical picture of light, Dirac assumed that light of every frequency can be described as a harmonic oscillator (with an analogy with a swing as discussed in Chapter 3). This is a justified picture as, at any point, the crest of the wave oscillates just like a swing or a harmonic oscillator. Treating light as a harmonic oscillator is a classical picture that can explain phenomena such as interference and diffraction.

Now according to Planck's hypothesis, the energy of a harmonic oscillator is quantized with each quantum or photon having an energy equal to hf, where h is Planck's constant and f is the frequency of the oscillator. Thus, a particle-type picture is associated with each color of light.

Dirac, in a truly ingenious manner, combined these two pictures in such a way that it could explain, on the one hand, phenomena such as interference and, on the other hand, phenomena such as the photoelectric effect in a unified manner. This was a great triumph of the quantum theory.

9.1 Quantum vacuum

There was an unexpected consequence of this simple but elegant picture of light. It turned out that, when absolutely no light is present, there is still some energy present for each color. This is highly counterintuitive. How can some finite energy be present associated with each frequency or color of light when there is no light?

Dirac showed that the amount of energy when no photon is present is equal to half the energy of each photon, $hf/2$. This means that, when one photon is present, the total energy is $hf + hf/2 = 3hf/2$, when two photons are present, total energy is $2hf + hf/2 = 5hf/2$, and so on. The surprising result is that, in vacuum, when no photon is present, there is still energy equal to $hf/2$ present. There are an infinite number of colors in the universe, each associated with a different frequency f. Thus, the total energy in the universe can be calculated by adding the vacuum energy for each color and the result is an infinite amount of energy. This is a startling result. The vacuum, which was considered devoid of any energy, has in fact an infinite amount of energy. The contrast could not be more glaring.

But this is not all.

According to the Heisenberg uncertainty relation, we cannot have a precise value of energy at a given time. The pair, energy and time, obeys the uncertainty relation, meaning that, for a very short time interval, there could be almost unlimited energy associated with every color, every frequency. The constant fluctuations of energy can produce massive particles not just out of thin air but out of absolutely nothing. The massive particles are always produced in pairs: particles and anti-particles such as an electron and a positron. These particles exist for a brief time and then disappear as energy. The presence of particles and anti-particles was first conjectured by Dirac and represents another amazing consequence of quantum mechanics. Since these particles have an extremely short life, they are called virtual. In most cases they cannot be observed directly or indirectly. The quantum theory therefore predicts that vacuum teems with virtual particles flitting in and out of existence.

We thus have a revolutionary new way of thinking about light that field quantization introduced into the scientific discourse, namely that light, when quantized, has the ability to exist in a state of pure nothingness—the so-called vacuum state—and yet have observable consequences in the material world.

The infinite energy in vacuum and the associated fluctuations are among the most counterintuitive aspects of the quantum world. These have some amazing consequences.

9.2 Spontaneous emission

Perhaps the most remarkable consequence of the fluctuations in the vacuum is the phenomenon of spontaneous emission by an atom. This effect is responsible for the light generated in conventional light sources like a light bulb. From a classical perspective, if an atom is in an excited state, it would remain in that state forever unless light of matching color is incident on it. However, the fluctuations of the light in vacuum can cause the atom to decay to the ground state and spontaneously emit a photon.

The process can be visualized as an apple on a tree. The apple will fall if a strong wind is blowing. It will not fall unless it is shaken up. If there is absolutely no wind, we would expect the apple to stay in the tree. Now if someone tells us that when there is absolutely no wind, there are still some fluctuations that will make the apple fall, we will look at that person with bewilderment.

The vacuum fluctuations provide such a mysterious source of disturbance that can get the electron in an excited state to be shaken up and fall to a lower state, releasing a photon in the process.

We can understand the process of light generation in a light bulb via spontaneous emission. A light bulb typically consists of a thin tungsten wire. At any time, most of the electrons inside the atom are in the lowest energy state. When an electric current flows through the tungsten wire, some atoms jump to the excited state. These atoms, without any stimulus from an external field, can decay to the ground state and emit light.

9.3 Lamb shift

A hydrogen atom consists of a single proton surrounded by an electron. This is one of the simplest systems in nature. The hydrogen atom is also a gift of nature as it is one of the extremely few real systems that can be studied without a computer and exact predictions can be made that can be tested in the laboratory. We have seen that Niels Bohr came up with a theoretical picture only for the hydrogen atom that could explain the observed colors of light by using the quantum conditions. When he tried to apply a similar model to more complicated systems like helium atoms, the model failed. This was when a need was felt to develop a theory that would not only reproduce the results for hydrogen atoms, but also give correct results for more complicated systems in a unified manner. When Heisenberg, and later Schrödinger and Dirac, developed the quantum theory, the first major test was to solve the hydrogen atom problem and explain the color structure of the emitted light.

When trust in quantum theory had been established in the 1930s, a major crisis arose. It was observed by Willis Lamb that there was an extremely small difference between the observed colors emitted by the hydrogen atom and the established equations of quantum theory. An experimental observation of this discrepancy became possible due to improvement in the measurement devices. After the second world war was over, in which many prominent physicists had played key roles in developing weapons of mass destruction, like the atom bomb and radar, the major problem facing physicists was to explain the Lamb shift of electronic levels inside the hydrogen atom. The question was what was missing from then-current quantum theory that failed to account for the deviation from the predicted energy levels.

It was soon realized that the vacuum fluctuations were responsible for the Lamb shift. The basic idea was that the electron inside the hydrogen atom is affected by the fluctuations associated with vacuum and when these fluctuations were included in the theory, the correct Lamb shift was obtained. The explanation of the Lamb shift is perhaps the greatest triumph of the field quantization. This gave rise to another field of physics called quantum electrodynamics.

How to understand the Lamb shift?

Here we present a heuristic picture. Instead of using a real quantum mechanical picture of electronic energy levels in the hydrogen atom, we consider the simpler Bohr model in which the electron can exist in certain well-defined orbits but not in any orbit in between. In the absence of the vacuum fluctuations, the orbits are well defined and, when the electron jumps from a higher orbit to a lower orbit, a photon of certain frequency is emitted. When the vacuum fluctuations are included, the electronic orbit is no longer well defined. Certain electronic orbits are shaken under the action of these fluctuations. The net result is that these orbits acquire a higher energy. When electrons jump from such orbits to lower orbits, the frequency, and hence the energy, of the emitted photon is slightly higher. This is the frequency shift that Willis Lamb observed.

9.4 Casimir force

In 1947, Hendrick Casimir predicted that if two conducting plates separated by a distance are placed in a vacuum, as shown in Fig. 9.1, and no external force is acting on them, they would attract each other. How can that be? How can the two metallic plates with nothing, absolutely nothing, in between them and no force of any kind acting on them attract each other? As we discuss below, the cause of this force is nothing but the infinite energy stored in vacuum. This is a highly counterintuitive result. One of the foremost scientists of the twentieth century, Julian Schwinger, called it "One of the least intuitive consequences of quantum electrodynamics," and according to another leading scientist, Bryce DeWitt, "What startled me, in addition to the crazy idea that a pair of electrically neutral conductors should attract one another, was the way in which Casimir said the force could be computed, namely, by examining the effect on the zero-point energy of the electromagnetic vacuum caused by the mere presence of the plates. I had always been taught that the zero-point energy of a quantized field was unphysical." The Casimir force was experimentally observed in 1958.

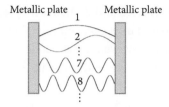

Metallic plate   Metallic plate

Fig. 9.1 Two metallic plates divide the vacuum in three regions. The outside regions have a continuum of infinite vacuum modes (not shown) and the region inside has an infinite number of discrete modes.

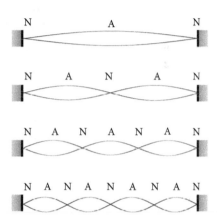

Fig. 9.2 Standing waves with (a) one anti-node, (b) two anti-nodes, (c) three anti-nodes, and (d) four anti-nodes. A stands for anti-node and N stands for node.

Casimir explained this force arising purely as a consequence of the infinite amount of energy in vacuum.

Before describing the Casimir force, we first explain how only some waves are allowed between two metallic plates separated by a distance L. As we see in Fig. 9.2, only those waves are allowed for which the peak amplitude oscillates in time but does not move in space. Such waves are called standing waves. In a standing wave, certain points on the wave, called "nodes," remain fixed but the points with non-zero amplitude, called "anti-nodes" oscillate at the same location with frequency f. For the distance L equal to half a millimeter, the lowest frequency wave that is allowed has a frequency of 300 billion cycles per second. The only other allowed frequencies are an integral multiple of this frequency. For example, the frequencies 600 billion cycles per second, 900 billion cycles per second etc. are allowed. However, any frequency in between, such as

400 billion cycles per second and 700 billion cycles per second, are not allowed. As the number of integers is infinite, there is an infinite number of allowed frequencies. However, these frequencies are discrete. In contrast, all possible frequencies are allowed in free space where there are no boundaries.

When the two conducting plates are inserted in the vacuum, the space is divided into three regions, the two infinite regions outside the plates and another region inside the two plates. The regions outside the plates have a continuum of frequencies, i.e., all possible frequencies, resulting in an infinite amount of energy when we add the contributions of all the frequencies. The region inside the plates, however, allows only a discrete number of frequencies as discussed above. These are also an infinite number of frequencies, each an integral multiple of a fundamental frequency. The total amount of the vacuum field energy between the plates is also infinite. Thus, we have an infinite amount of energy outside the plates and an infinite amount of energy between the plates. The truly dramatic result is that when we subtract these two infinities, the outcome is finite. As the system tends to evolve to a state with minimum energy there is a resulting force and this force is attractive.

How can one infinity be larger than another infinity? If this is true, it would be strange, to say the least. In order to prove this result, let us consider two infinities, one consisting of integers from 1 to infinity listed in the first column and all the numbers between 0 and 1 listed in the second column:

1	$0.7932462\cdots$
2	$0.5374291\cdots$
3	$0.6279812\cdots$
4	$0.5712460\cdots$
5	$0.1435452\cdots$
6	$0.8790421\cdots$
7	$0.3572487\cdots$
$\vdots$	$\vdots$

To say that the infinite numbers between 0 and 1 are larger than the infinite number of integers defies common sense. After all, if we come up with any number between 0 and 1 in the second column, it can be identified with an integer as there are an infinite number of integers. However, a closer look indicates that it is plausible that one infinity is larger than the other. To see this, we note that the difference between the two infinities is that, one is discrete (integers can be 1, 2, 3, ⋯ ⋯) and the other is continuous (every possible number between 0 and 1). Another way of looking at it is that, between two integers (say 5 and 8) there are a finite number of integers (6 and 7). However, between two numbers between 0 and 1, no matter how close they are, there are still an infinite number of numbers.

Now comes the proof that indeed the infinite number of numbers between 0 and 1 is larger than an infinite number of integers. It is essentially a one-line proof. Let us construct a number between 0 and 1 by adding 1 to the first digit of the first number, the second digit of the second number, the third digit of the third number, and so on. The resulting number is 0.8483538⋯. However, this number cannot be the first number as, at least the first digit is different. It cannot be the second number as the second digit will be different, And so on. This shows that at least one number is not on the list of infinite numbers between 0 and 1 that matches each integer on the infinite numbers of integers. This proves that the infinite number of numbers between 0 and 1 are larger than the infinite number of integers. This is a remarkable result.

It is now simple to understand why the total energy outside the plates is larger than the energy between the plates, even when both energies are infinite. The infinite energy between the plates comes by adding a discrete number of frequencies associated with the integers whereas the infinite energy outside the plates is continuous and corresponds to all possible frequencies. The remarkable result is that when the two infinities are subtracted, we get a finite number. This explains the Casimir force between the two plates and the force arises only due to the infinite amount of vacuum energy.

9.5 Can we extract photons from the vacuum?

A question of interest is whether we can extract real photons from this sea of infinite vacuum energy. This would be akin to a magician producing a rabbit out of a hat. If this becomes possible, our energy problem will be over forever. We would be able to run machines, all kind of machines, using this infinite source of energy.

However, nature does not allow us to have things for free—there is a price to be paid for everything. According to the laws of physics known to us, it is impossible to build a perpetual motion machine. The situation of using the sea of vacuum energy to run a machine, like a car, is akin to a ship travelling in an ocean of warm water. The ocean water has also an almost unlimited supply of heat energy. But the ship cannot run on this energy alone. The laws of thermodynamics do not allow the extraction of energy from the ocean water with no external source of energy.

The laws of thermodynamics have proven to be amazingly resilient. The foundations of Newtonian mechanics and electromagnetic theories crumbled with the advent of the revolutionary theories of relativity and quantum mechanics at the beginning of the twentieth century. However, surprisingly, the laws of thermodynamics survived these new ideas about light and matter. The second law of thermodynamics remains the stumbling block in taking advantage of this sea of infinite energy that we call vacuum.

There are, however, other possibilities where the virtual photons in vacuum can be converted into real photons. However, they require an external source of energy to do so. Here we mention two of them.

First, we consider the same set-up as for the Casimir force: two metallic plates held parallel to each other. It was shown by Gerald T. Moore in 1969 that, if these metallic plates vibrate very fast then the virtual photons created by vacuum fluctuations can be converted into directly observable real photons. This effect, which has been observed experimentally is called the dynamic Casimir effect (Fig. 9.3).

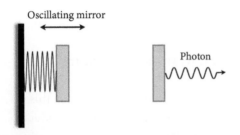

Oscillating mirror

Photon

Fig. 9.3 Schematics for the dynamic Casimir effect. One metallic plate is kept fixed whereas the other plate oscillates with very high frequency.

Another possibility is via the Unruh effect. William Unruh showed that the notion of vacuum depends on the motion of the observer. If an object accelerates, it would perceive the vacuum (with no photons) as having acquired some photons. The energy of these photons would depend on the acceleration.

Another way of looking at the Unruh effect is that the complete vacuum exists when the temperature is almost absolute zero (0 Kelvin). At this temperature, there are no photons present of any frequency. However, if an object like a thermometer is accelerated through this vacuum, it would exhibit a non-zero temperature. The temperature is proportional to the acceleration. The accelerated particle sees not vacuum but a radiation field with characteristics like heat radiation. This radiation, called Unruh radiation, can excite the electrons inside an atom to higher energy levels, thereby converting the vacuum energy into real useful energy. This effect has not been observed yet due to the large acceleration required. An acceleration of 10^{17} kilometers per second per second is required to see a rise in temperature of 1 degree Kelvin.

10

Wave–particle duality

These experiments (Scully's quantum eraser) are a magnificent affront to our conventional notions of space and time. Something that takes place long after and far away from something else nevertheless is vital to our description of that something else. By any classical-common sense-reckoning, that's, well, crazy. Of course, that's the point: classical reckoning is the wrong kind of reckoning to use in a quantum universe. For a few days after I learned of these experiments, I remember feeling elated. I felt I'd been given a glimpse into a veiled side of reality.

— Brian Greene

Our common sense is based on what we observe in the world around us. In our observations, particles and waves are distinguishable phenomena, with different characteristic properties and behaviors. Particles are massive objects that can occupy well-defined positions and can move with well-defined speed. They can collide with each other. We cannot associate any of these properties with waves. Waves are traveling disturbances and are found in an amazingly diverse range of physical systems. The concept of a wave is most clearly understood by looking at the periodic disturbances through the water when we drop a small object like a pebble in a body of still water. The disturbance produces water waves, which move away from the point where the pebble entered the water. Examples of particles are a tennis ball, a dust particle, an atom, and an electron and the most well-known examples of waves are light and sound.

An important property of any kind of wave is that, when two or more waves arrive at the same point, they superimpose themselves upon one another. This is called the principle of superposition. As simple examples, we consider two waves in Fig. 10.1a and Fig. 10.1b. In Fig. 10.1a, the

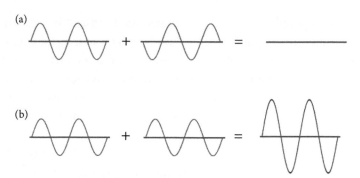

Fig. 10.1 Principle of superposition: Two waves when, at each point: (a) they have equal and opposite amplitude resulting in destructive interference with a wave of zero amplitude; (b) they have equal and the same amplitude resulting in constructive interference with a wave of twice the amplitude.

two waves are shifted with respect to each other. According to the principle of superposition, both waves cancel each other at every point and the resulting *amplitude* at each point is zero. In Fig. 10.1b, the two waves have the same amplitude at each point and the resulting wave is identical to each wave with the only difference being that the *amplitude* is twice of each individual wave. We say that the two waves interfere destructively in Fig. 10.1a and constructively in Fig. 10.1b. This important principle has many applications, particularly in understanding interference and diffraction.

It is a common observation that a particle like a ball or an electron never behaves like a wave and a wave like light or sound never behaves like a particle. This is the kind of world we seem to live in. The behaviors of both particles and waves were fully understood within the framework of classical mechanics at the end of the nineteenth century.

In quantum mechanics, such a distinction is done away with. A counterintuitive result is that both light and matter display dual behaviors. In some situations, they behave like waves and in others they behave like particles. For example, the interference and diffraction phenomena can only be explained by treating light as a wave. These phenomena cannot be explained by treating light as consisting of particles. On the other hand, Einstein explained the photoelectric effect in 1905 by treating light as composed of quanta or photons. As discussed in Section 3.2, the wave picture of light fails to explain how light of a certain color, no matter how intense, cannot eject an electron from the metallic surface, yet light of another color, no matter how feeble, can cause an electron to be emitted. In 1924, the other side of wave–particle duality was argued by de Broglie when he showed that a massive object can behave both as a particle and a wave.

A mysterious consequence of quantum mechanics is that the choice whether light or matter behave as wave or particle depends on the experimental set-up. The quintessential experiment that demonstrates wave–particle duality is Young's double-slit interference experiment. We elaborate wave–particle duality in nature via this historically important experiment for light as well as for electrons.

10.1 Young's double-slit experiment for waves and particles

In Young's original double-slit experiment, a light beam was passed through two slits as shown in Fig. 10.2c. This resulted in an interference pattern, a pattern of bright fringes separated by dark fringes, on a screen, as shown in Fig. 10.2c. The reason for such a pattern is that, except at the mid-point, the light at any point on the screen from the two slits travels different distances. The bright fringes are located at those points on the screen where the path difference between the light waves from the two slits is such that they interfere constructively. The dark fringes are, on the other hand, located at those points where the two waves interfere destructively. Central to this description is the wave nature of light. We also note that, if the lower slit is blocked as in Fig. 10.2a or the upper slit is blocked as in Fig. 10.2b, the interference pattern disappears and we get a bright spot in the upper half and the lower half, respectively.

We can understand this behavior with the help of Fig. 10.3. Light from a source is incident on a screen that contains two narrow slits. We look at the amount of light at the screen at a point P. We notice that the distance light waves travel from the lower slit are longer than the light waves that travel from the upper slit. The difference in the two distances determine whether there will be a brightness or a darkness at point P. If the path

(a)

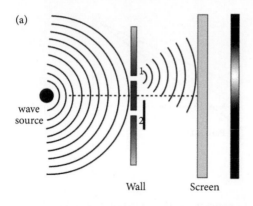

(b)

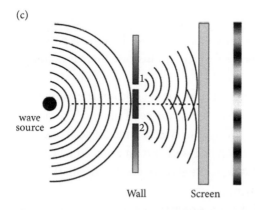

(c)

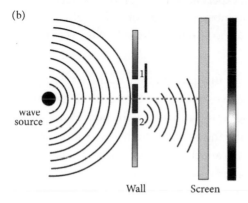

Fig. 10.2 Young's double-slit experiment with waves. (a) When the lower slit is blocked, there is a bright spot in the upper part of the screen, (b) when the upper slit is blocked, there is a bright spot in the lower part of the screen, and (c) when both slits are open, an interference pattern consisting of bright and dark spots is obtained.

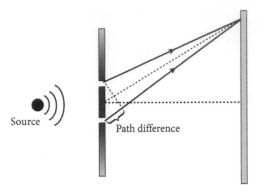

Fig. 10.3 In Young's double-slit experiment, the point on the screen is bright if the path difference is an integral multiple of wavelength and is dark if the path difference is equal to odd multiples of half the wavelength.

difference is equal to zero or an integral multiple of the wavelength of the light, constructive interference takes place between the light waves coming from the two slits and a bright fringe is formed. However, for those points on the screen where the waves from the two slits are shifted like in Fig. 10.1a, the two waves cancel each other out and a dark fringe is obtained. An interference pattern is formed. It is clear that the appearance of the interference pattern (a pattern of dark and bright fringes) is possible only because light acts like a wave.

In the case where the upper or the lower slit is blocked, as shown in Figs 10.2a and 10.2b, there is only one wave that hits the screen—the result is that a single bright spot is formed. In this case there is no longer interference.

Next, we consider the double-slit experiment with particles like bullets as shown in Fig. 10.3. Here a gun is a source of bullets. The bullets, sent in the forward direction, are spread over a wide angle. These bullets can pass through holes 1 and 2 in a wall and hit a screen where they are detected. Unlike the light waves, there is no interference in this case. What we observe is the following.

When hole 2 is covered, bullets pass only through slit 1. The probability of a single bullet hitting the screen is in the upper part of the screen. The maximum probability occurs at a location which is on a straight line with the gun and slit 1. When a large number of bullets are incident

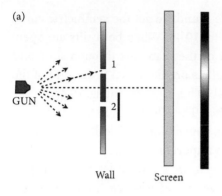

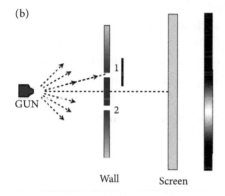

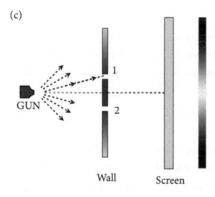

Fig. 10.4 Young's double-slit experiment with bullets. (a) When the lower slit is blocked, the bullets hit the upper half of the screen, (b) when the upper slit is blocked, the bullets hit the lower half of the screen, and (c) when both the slits are open, the result is the sum of the two distributions (a) and (b). There is no interference in this case.

on the screen, their distribution (fraction of the total number of bullets hitting the screen) is shown in Fig. 10.4a. This curve is identical to the intensity distribution for the waves in Fig. 10.2a. When slit 1 is closed, bullets can only pass through hole 2 and we get the symmetric curve for the distribution as shown in Fig. 10.4b. When both slits are open, the bullets can pass through slit 1 or they can pass through slit 2 and the resulting distribution of the bullets on the screen is the sum of the distributions in Figs 10.4a and 10.4b. This is shown in Fig. 10.4c.

The probabilities just add together. The effect with both slits open is the sum of the effects with each slit open alone. We call this result an observation of "*no interference*." An important point to note here is that, for each bullet detected on the screen, we know (at least in principle) which slit it came from, i.e., we have the "*which-path*" information for each bullet. Indeed, we can determine the full trajectory of each bullet from the point it leaves the gun and hits the screen.

So far, we have considered Young's double-slit experiment with waves and with bullets. In the case of waves, we observe interference. However, when we repeat the same experiment with bullets, we find no interference.

What about Young's double-slit experiment with electrons? Do electrons behave like bullets or do they behave like waves?

10.2 Young's double-slit experiment with electrons

We consider electrons being emitted by an electron gun. This beam of electrons passes through a wall with two slits as shown in Fig. 10.5. The set-up is identical to the set-up for the double-slit experiment for bullets. But do we get the same result as those for the bullets?

When slit 2 is closed, electrons can only pass through slit 1. The probability of a single electron hitting the screen is, as for the bullets, in the upper part of the screen. Similarly, the probability of a single electron is symmetrically in the lower part of the screen when slit 1 is closed and the electron can pass through slit 2 only. These curves are identical to the corresponding curves when the bullets are incident on the screen and also identical to the intensity distribution when a beam of light is incident.

But what happens when both slits are open? Do electrons behave like particles as bullets or do they behave like waves as a light beam? The

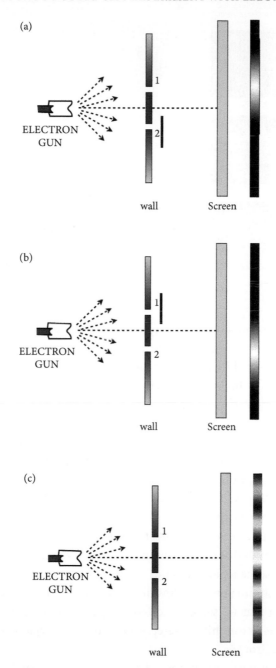

Fig. 10.5 Young's double-slit experiment with electrons.

(a)

(b)

(c)

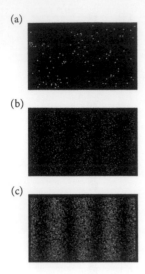

Fig. 10.6 The outcome of Young's double-slit experiment with (a) 100 electrons, (b) 1000 electrons, and (c) 10,000 electrons.

results of the double-slit experiment in which electrons are incident on the two slits one by one are shown in Fig. 10.6. Each electron after passing through the slits hits the screen and is detected at different locations. Here we see the build-up of electrons on the screen. For 100 electrons, the distribution of the detected electrons on the screen appears to be random. After about 1000 electrons are detected, the distribution on the screen seems to have some regions with a dense distribution compared to others. But, still it is difficult to conclude anything regarding the particle or wave behavior of the electrons.

After 10,000 electrons are detected on the screen, there is an unmistakable interference pattern with bright fringes, separated by dark fringes. The individual electrons are detected one by one, but instead of a pattern that is similar to that corresponding to bullets, we find the electrons are detected in some regions and not in others.

This is a stunning result. How did the electrons know where to hit the screen such that we see an interference pattern emerging after a large number of electrons hit the screen? Why are there regions where we do not find electrons?

This experiment was proposed by Richard Feynman in his famous Feynman Lectures in 1965 in these words:

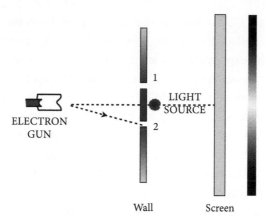

Fig. 10.7 Young's double-slit experiment with which-path information. A photon from the light source scatters from an electron and provides the which-path information. The which-path information leads to the disappearance of the interference fringes.

> We choose to examine a phenomenon which is impossible, *absolutely* impossible, to explain in any classical way, and which has in it the heart of quantum mechanics.

He, however, claimed that the experiment is too difficult to carry out and may never be done. What Feynman apparently did not know was that a double-slit experiment with electrons had already been done by Claus Jönsson in 1961.

As mentioned earlier (Section 4.5), Young's double-slit experiment applied to the interference of electrons was voted as the Most Beautiful Experiment in Physics in a survey in 2002.

The situation becomes more mysterious when a slight variation of this experiment gives us a completely different outcome.

Let us place a source of light between the two slits as shown in Fig. 10.7. When an electron passes the slits, light scatters from the electron and provides us with the which-path information. We now know for each electron which slit it passed through to get to the screen. After a large number of electrons are detected at various points on the screen, the resulting pattern is the same as that obtained in the double-slit experiment for the bullets. In this case, the interference disappears. This is in contrast to the experiment depicted in Fig. 10.5c, where we had a lack of knowledge about the path each individual electron took. This lack of which-path knowledge seems to be responsible for interference.

Thus, if we "look" at *which path* each electron followed, the interference disappears and we get the same distribution on the screen as for the particles. The result is that either we get interference when we have *no which-path* information or we lose interference when we have the *which-path* information.

No classical explanation can describe these observations—we can reconcile these observations only with the fundamental principles of quantum mechanics.

We describe the electron, not as a particle traveling in a well-defined trajectory, but by a wavefunction ψ which is a function of position. At any point R on the screen, there are two contributions for the same electron coming from the two slits, ψ_1 and ψ_2.

When slit 2 is closed, the total wavefunction at the position R is ψ_1 and the probability of finding the electron is $P_1 = |\psi_1|^2$. Similarly, when slit 1 is closed, the total wavefunction at the position R is $\psi_2(R)$ and the probability of finding the electron is $P_2 = |\psi_2|^2$. When both slits are open, the total wavefunction of the electron at the position R is $\psi = \psi_1 + \psi_2$. At those points on the screen where $\psi_1 = \psi_2$, there is constructive interference and a larger number of electrons are detected. Similarly, at those points on the screen where $\psi_1 = -\psi_2$, the total wavefunction $\psi = 0$, and there is destructive interference and no electron is detected at those locations. Thus a pattern similar to the interference pattern for light is obtained.

If an experiment is performed which is capable of determining whether the electrons passed through slit 1 or slit 2, the probability of finding the electron at a point R on the screen is the sum of the probabilities for each alternative and the interference is lost.

This concept of wave–particle duality has been a source of intense discussion since the earliest days of quantum mechanics. How the same electron can behave like a wave in one situation and a particle in another is quite mysterious.

Richard Feynman expresses his amazement at these incredible results in these words:

> One might still like to ask: "How does it work? What is the machinery behind the law?" No one has found any machinery behind the law. No one can "explain" any more than we have just "explained." No one will give you any deeper representation of the situation. We have no ideas about a more basic mechanism from which these results can be deduced.

Here we have discussed the experiment with electrons. The same can be said about a similar experiment with light. If we treat the light beam in Young's double-slit experiment as consisting of a large number of photons, the situation is similar to the interference experiment with electrons. The reason we get interference of a light beam in a double-slit experiment is due to the lack of which-path information for each photon. If somehow we are able to get the which-path information for each photon, the interference disappears.

This is the essence of wave–particle duality or Bohr's principle of complementarity: Electrons and photons can behave like waves when we have no which-path information and they behave like particles when we have the which–path information.

Wave–particle duality was a subject of a fierce debate between Albert Einstein and Niels Bohr, as we discuss next.

10.3 Einstein–Bohr debate

Wave–particle duality is a manifestation of Niels Bohr's principle of complementarity. According to Bohr, in any quantum mechanical experiment, certain physical concepts are complementary. If the experiment clearly illustrates one concept the other concept will be completely obscured. This implies that, if the particle nature of an object is exhibited in an experiment, then the wave nature will be completely obscured.

Thus, in the two-slit experiment, we can either have the *which-path* information or the existence of an interference pattern. According to

Bohr's principle of complementarity, they can never be observed at the same time, in the same experiment.

Einstein, however, came up with a clever scheme such that we can have both the which-path information (particle nature) and the interference (wave nature) in the same experiment—violating Bohr's principle of complementarity.

In Einstein's proposed experiment a wall with the two slits is placed on rollers so that it can move freely in the vertical direction as shown in Fig. 10.8. An electron gun shoots electrons toward the wall where they can pass through the two slits and then onto the back screen to create the interference pattern.

The electrons move with a certain speed in the forward direction. However the electron beam is spread out such that some electrons pass through slit 1, some pass through slit 2, and the rest collide with the wall and do not contribute to the interference. The question is whether the electrons passing through the two slits form an interference pattern as discussed earlier, but at the same time, we find out which slit each electron passed through. Einstein argued that, for each electron, we can find out which slit it passed through by measuring the movement of the wall

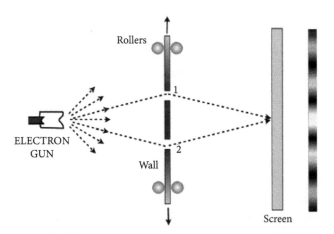

Fig. 10.8 Einstein's *gedanken* experiment. Electrons pass through a narrow slit through a wall that can freely move on a roller. The momentum transfer to the wall provides the which-path information. At the same time electrons from the two slits can form an interference pattern on the screen.

in the vertical direction. If the wall moves up, we can conclude that the electron passed through slit 1 and if it moves in the downward direction, the electron must have passed through slit 2. Therefore, by observing the motion of the wall (upward or downward), the which-path information of each electron is obtained. The important point is that this information is obtained without disturbing the electron.

Einstein then argued that this modified set-up can simultaneously exhibit both wave and particle characteristics in a simple experiment. According to him, the situation is very similar to the double-slit experiment discussed in the last section.

After passage through the slits, the undisturbed electrons can proceed to the screen and give the interference pattern as before. However, we can get the information about which slit the electrons passed through by measuring the movement of the wall after each electron has passed through. Thus we have both the "which-path" information and the interference. This is in contradiction to Bohr's principle of complementarity.

This was a forceful argument against the foundational principle of quantum mechanics and Bohr had to respond to it immediately. Leon Rosenfeld records the encounter, that took place at the 14th Solvay Conference, in the following words[1]:

... Einstein thought he had found a counterexample to the uncertainty principle. It was quite a shock for Bohr ... he did not see the solution at once. During the whole evening he was extremely unhappy, going from one to the other and trying to persuade them that it couldn't be true, that it would be the end of physics if Einstein were right; but he couldn't produce any refutation. I shall never forget the vision of the two antagonists leaving the club [of the Fondation Universitaire]: Einstein a tall majestic figure, walking quietly, with a somewhat ironical smile, and Bohr trotting near him, very excited ... The next morning came Bohr's triumph.

Bohr invoked the Heisenberg uncertainty relation to refute Einstein's argument and saved the principle of complementarity.

---

[1] *Elementary Particle Physics* (Proceedings of the Fourteenth Solvay Conference, Interscience, New York, p. 232).

According to the Heisenberg uncertainty relation, if we determine the speed of an object in a certain direction very precisely, we cannot, at the same time, know its position very accurately. The more certain we are about the speed of the object, the less certain we are about its location. This was discussed in Section 7.4. In Einstein's argument, it is necessary to know the speed of the wall before the electron passes through it sufficiently precisely. This is required as we need to know the change in the speed of the wall in the vertical direction after the electron has passed in order to obtain the which-path information. However, according to the Heisenberg uncertainty principle, we cannot know the position of the wall in the vertical direction with arbitrary accuracy. Therefore, a precise measurement of the speed means that the locations of the slits become indeterminate. The uncertainty in the location of the slits means that the electrons effectively see a blurred pair of slits. The locations where electrons hit the screen consequently become random and the center of the interference pattern has a different location for each electron, thus wiping out the interference pattern. This shows that the which-path information in the Young's double-slit experiment smears the interference pattern.

10.4 Delayed choice and quantum eraser

The Young's double-slit experiment becomes even more mysterious when we include delayed choice in the experimental set-up.

In Young's double-slit experiment, whether we get the interference fringes or not depends on whether we have no which-path information or we have the which-path information. Thus a photon behaves like a wave or a particle depending upon what kind of an experiment we decide to do. If we decide not to look at the photon when it is passing through the slits, it behaves like a wave. However if we decide to find which slit the photon goes through, it behaves like a particle. This wave–particle duality is very mysterious and it becomes even more so when we try to address the question as to whether the photon knew in advance what behavior it should exhibit. This question was addressed by John Wheeler in his "delayed choice" *gedanken* experiment.

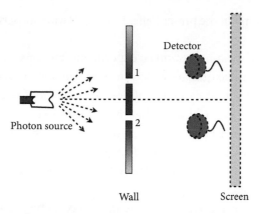

Fig. 10.9 A modified scheme for Young's double-slit experiment. A stream of single photons is generated by the source. The decision whether to place the screen or the detectors is taken after the photon has passed through the wall with the two slits.

Let us consider a modified double-slit experiment as shown in Fig. 10.9. Here, light after passing through the two slits in the wall propagates toward the screen. We consider the system in which a stream of single photons is incident on the two slits. However, instead of a screen we have a choice of either having a screen or a pair of detectors. If there is a screen then, as seen before, after a passage of a sufficient number of photons, there will an interference pattern on the screen. Again, as discussed before, this can happen only if the photons passed through both the slits, exhibiting the wave nature of light. However, if there is a pair of detectors, either the upper detector will register the photon indicating that the photon passed through the upper slit or the lower detector will register the photon indicating that the photon passed the lower slit. In this case, we have the which-path information for each photon and there is no interference pattern. Light exhibits the nature of a particle.

The new twist is that one can decide whether to place the screen or the detectors *after* each photon passed through slits in the wall. This can, for example, be done by using very fast switching devices. In this "delayed choice" experiment, the photon decides *after* passing the two slits whether it passed through slit 1 (or slit 2) or passed through both the slits *in the past*. Did the choice of the experimental apparatus (a screen or the detectors) in the present affect the past (wave or particle nature of light)?

In Wheeler's original *gedanken* experiment, photons are generated by cosmic objects like quasars which may be trillions of miles away from earth. These photons can take millions or billions of years to travel to earth. They can be so far away that the photons were generated at those stellar objects well before earth, or the solar system, existed. Here we note that, as discussed in Section 18.5, the galaxies were formed over 13 billion years ago whereas our solar system came into existence about 4.6 billion years ago.

The photons can travel to earth with a galaxy acting as a gravitational lens as shown in Fig. 10.10. How a massive galaxy can behave like a gravitational lens is a consequence of Einstein's general theory of relativity and will be discussed in Section 16.6. Photons generated by distant quasars can then follow two different paths to travel to earth: a path left of the galaxy or the other right of the galaxy. After having traveled a distance of trillions of miles the photons arrive at earth where they are detected in one of the two different experimental set-ups.

In the first set-up, we place two detectors D_1 and D_2 as shown in Fig. 10.10. If the detector D_1 clicks, we can conclude that the photon followed the left path and if the detector D_2 clicks the photon followed the right

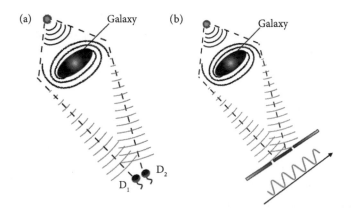

Fig. 10.10 Wheeler's delayed choice experiment. Light that left a quasar millions of years ago can be made to act (a) like a particle or (a) a wave depending on our choice of the experimental set-up.

path. Thus a click at either D_1 or D_2 provides the which-path information. For example, for a click at D_1, we can conclude that the photon was in the left path all along for all those billions of years. Similarly, for a click at D_2, we can conclude that the photon was in the right path all along.

The other possibility is to pass them through the two slits in a Young's double-slit experiment. These photons give rise to an interference pattern. We can then conclude that the photons behaved like waves and they went through both ways around the galaxy.

Therefore, in the first case the photons appear to pass through only one side of the galaxy and behave like particles and in the latter case they behave like waves and go through both ways around the galaxy. The paradoxical situation is that it depends on the experimenter's "delayed choice" whether the photon generated billions of years ago behaves like a particle or a wave. Until the experiment is done, we cannot say whether the photon will behave as a particle or as a wave.

This is astonishing: How could the measurement device (individual detectors or the screen with double slits) placed long after the photon was well on its way decide whether it travelled all those billions of years through one path and behaved like a particle or through both paths and behaved like a wave? Similar, delayed choice experiments have been done in laboratories and the results are all in agreement with the predictions of quantum mechanics.

One way out of the conundrum is to say that our present changes the past, that is, the experimental set-up now or in the future can change the past. If we have the set-up to detect the photons, then the photon changes its past accordingly by travelling along the right path or the left path. And if we put the interference set-up in place, the photon goes to the moment of its generation several years ago and travels through both paths all those years. But this mind-boggling scenario is too radical a resolution.

What is more reasonable (and this is also extremely mystifying) is to say that the photon paths or even photons have no reality until we decide to measure them. It makes no sense to say which path or paths the light photons took until an experiment is set up to look at them.

This simple but mysterious example illustrates in a forceful manner that our conventional concept of reality may be highly flawed. Reality seems to come into existence only when we try to observe. Without observation there is no reality. We shall examine the question of reality in the next chapter.

An even more counterintuitive aspect of wave–particle duality is the notion of a "quantum eraser" introduced by Marlan Scully and Kai Drühl in 1982. As discussed earlier, in a Young's double-slit experiment, we get an interference pattern if we have no knowledge about which slit the photon went through. However, if we somehow obtain the which-path information then the interference is lost. Scully posed the question: Is it possible to "erase" the which-path information and recover the interference pattern *after* the photon has passed through the slits and is detected on the screen?

We present a simple description of the quantum eraser as depicted in Fig. 10.11. The set-up is not much different from that of the Young's double-slit experiment. The main difference is that, in each slit, two photons are produced simultaneously instead of one. This can be done by inserting some special crystals that can generate either two photons, *a* and *b*, in the upper slit or two photons, *a* and *b*, in the lower slit. The *a* photons interfere just like in the Young's double-slit experiment. The role of the *b* photons is to provide the which-path information.

What about the appearance and disappearance of interference fringes discussed above?

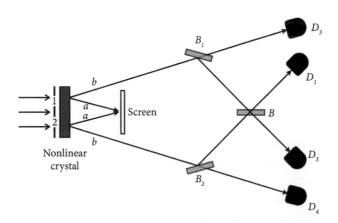

Fig. 10.11 Schematics of the quantum eraser.

For this purpose, we look at the a photon that proceeds to the screen. We consider only those instances where the b photon scattered from the slit located at 1 proceeds to the beam splitter B_1 and the b photon scattered from the slit located at 2 proceeds to B_2. At either of these 50/50 beam splitters, the b photon has a 50 percent probability of proceeding to detectors D_3 (for photons scattered from 1) and to D_4 (for photons scattered from 2). On the other hand, there is also a 50 percent probability that the photon will be reflected from the respective beam splitter and proceed to another 50/50 beam splitter, B. For these photons, there is an equal probability of being detected at detectors D_1 and D_2.

If the b photon is detected at detector D_3, it has necessarily come from the slit located at 1 and could not have come from the slit located at 2. Similarly, detection at D_4 means that the b photon came from the slit located at 2. For such events, we can also conclude that the corresponding a photon was also scattered from the same slit. That is, we have "'which-way'" information if detectors D_3 or D_4 register a count.

Returning to the quantum erasure protocol, if the b photon is detected at D_1, there is an equal probability that it may have come from the slit located at 1, following the path $1B_1BD_1$, or it may have come from the slit located at 2, following the path $2B_2BD_1$. Thus, we have erased the information about which slit scattered the a photon, and there is no which-path information available for the corresponding b photon. The same can be said about the b photon detected at D_2. After this experiment is done a large number of times, we shall have roughly 25 percent of b photons detected each at D_1, D_2, D_3, and D_4 because of the 50/50 nature of our beam splitters. The corresponding spatial distribution of a photons will be, as mentioned above, completely random.

Next, we do a sorting process. We separate out all the events where the b photons are detected at D_1, D_2, D_3, and D_4. For these four groups of events, we locate the positions of the detected a photons on the screen D. The key result is that, for the events corresponding to the detection of b photons at detectors D_3 and D_4, the pattern obtained by the a photons on the screen D is the same as we would expect if these photons had scattered from slits at sites 1 and 2, respectively. That is, there are no interference fringes, as would be expected when we have which-path information available. On the contrary, we obtain interference fringes for those events where the b photons are detected at D_1 and D_2. For this set of data, there is no which-path information available for the corresponding a photons.

Suppose we place the b photon detectors far away. Then the future measurements on these photons influence the way we think about the a photons measured today (or yesterday!). For example, we can conclude that a photons whose b partners were successfully used to ascertain which-path information can be described as having (in the past) originated from site 1 or site 2. We can also conclude that a photons whose b partners had their which-path information erased cannot be described as having (in the past) originated from site 1 or site 2 but must be described, in the same sense, as having come from both sites. The future helps shape the story we tell of the past.

The quantum eraser and its experimental realization bring out the counterintuitive aspects related to time in the quantum mechanical domain.

10.5 Interaction-free measurement

Quantum interference, as embodied in the Young's double-slit experiment, brings out the amazing consequences of wave–particle duality. In this and the next section, we show how wave–particle duality manifested in the no which-path and the which-path information in another interference set-up can lead to some novel and amazing phenomena.

It is a common observation that the presence of an object, any object, can be inferred by a direct observation or interaction with the object itself. For example, we can see all the things around us due to the light that scatters from the objects into our eyes. In general, the act of "looking" implies that something like light scatters from the object and is registered in the measuring device (such as our eyes). If there is no light (visible or invisible), the presence of the objects cannot be inferred. Is it possible to "observe" an object without ever looking at it? Conventional wisdom tells us that this should not be possible. However, the mysterious quantum mechanical properties allow such a counterintuitive possibility—we can see without looking. This happens as a result of a method called interaction-free measurement.

The concept of interaction-free measurement was first suggested by A. C. Elitzur and L. Vaidman in 1993. In a colorful example, they proposed it as a method for bomb testing. Suppose we have a collection of bombs, some good and some bad. The bad bombs do not absorb photons and let them pass through unhindered. The good bombs however absorb the photons and explode. This makes it impossible to identify a good bomb without exploding it by shining light on it and measuring the scattered light. In any conventional method, the good bombs will always explode in an attempt to find them. The question is: Can we make an optical measurement to find at least some of the good bombs without exploding them?

We consider an interferometer with two beam splitters and two mirrors as shown in Fig. 10.12a. This is called a Mach–Zehnder interferometer. The input is a single photon in one port of the beam splitter and none at the other. The role of the beam splitter is to either reflect the incident photon to the left side or to transmit it to the right side. This is a probabilistic event. We consider two situations: when the object is not

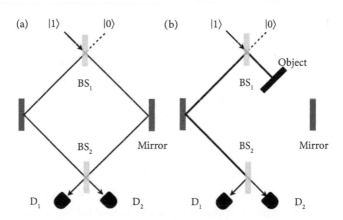

Fig. 10.12 (a) In a Mach–Zehnder interferometer, a photon incident on a beam splitter BS_1 is reflected by mirrors, recombined at the second beam splitter BS_2, and is detected either at detector D_1 or D_2. (b) The same set-up as in (a) but with an object on the right-hand side of the interferometer.

present (Fig. 10.12a) and when the object is present (Fig. 10.12b) in the right arm of the interferometer.

In the absence of an object (Fig. 10.12a), the photon, like in the double-slit experiment, is in a *superposition of states* of being on the right side as well as on the left side after passing the first beam splitter. The role of the mirror is to reflect the incident photon. A result of quantum interference is that, after the passage through the second beam splitter, the quantum amplitudes add constructively on the left side and destructively on the right side. The result is that the photon is detected with unit probability (certainty) by the detector D_1.

However, if an object is present as shown in Fig. 10.12b, the photon, if transmitted to the right side by the beam-splitter BS_1, will be absorbed by the object. In this case, no photon will be detected either by the detector D_1 or the detector D_2. If the object is a bomb of the type discussed above, it would explode. However, if the photon is reflected on the left side, it would arrive at the beam-splitter BS_2 from the left side. There is no interference in this case and, after passing through BS_2, there is an equal probability that the photon will be detected at one of the detectors D_1 or D_2.

At this point, we have a counterintuitive result. If the detector D_1 clicks, we do not know whether the photon passed through both left and right paths (no which-path information) or it passed through the left path (which-path information). We, therefore, cannot conclude anything about the presence or absence of the object on the right side. However, in those cases that the detector D_2 clicks, we can be certain that the photon travelled through the left path AND the object was present on the right path. In those cases, the object is detected without the photon being directly incident on the object. Thus, an *interaction-free measurement* is obtained.

In the simple set-up shown in Fig. 10.12, the success probability is only 25 percent. To see this, we note that, for 50/50 beam splitters, there is a 50 percent probability that the photon would hit the object. For the remaining 50 percent, there are equal probabilities (25 percent each) for the detectors D_1 and D_2 clicking. The successful event is when detector D_2 clicks, which happens with 25 percent probability. This probability can be increased via more complicated set-ups.

10.6 Counterfactual quantum communication

It has always been a self-evident feature of any kind of communication that there should be an exchange of objects between the sender and the receiver to convey any information. For example, light pulses or photons are carriers of information in optical communication. For a binary information transfer (0 or 1), the information can be coded—a horizontally polarized photon as "0," and a vertically polarized photon can be coded as "1." Thus, if, in a communication between (say) Alice and Bob, Alice wants to send "0," she sends a horizontally polarized photon and if she wants to send a "1," she sends a vertically polarized photon. Another example is that Alice may decide to send no photon if she wants to send "0" and send a photon if she wants to send a "1." In these and any other examples of communication, an exchange of an object, like a photon, always takes place even if it is for one half of the communication (as in the second example). It is inconceivable that a communication between Alice and Bob can take place in such a way that there is no photon present in the channel of transmission (such as empty space or fiber-optic cables) in both cases, when the sender wants to send a "0" or wants to send a "1."

It was shown in 2013 by my research group that quantum mechanics can be used to achieve this apparently impossible objective. Such a communication protocol is called counterfactual communication. Counterfactual communication is thus optical communication with invisible photons. The protocol for counterfactual communication is schematically shown in Fig. 10.13.

The objective is for Bob to send binary information, "0" or "1," to Alice. As shown in Fig. 10.13a, Alice has a source of single photons, an optical set-up involving optical devices such as mirrors and beam splitters (represented by the box) as well as two photon detectors D_1 and D_2. Bob only has a mirror. Alice sends a photon to Bob such that it interacts with optical devices inside the box in a specified manner before proceeding to Bob via a transmission channel that can be free space or optical fiber. Bob has now two choices: He can either block the photon from reaching the mirror, for example, by placing his hand in front of the mirror (Fig. 10.13b) or he does nothing and allows the photon to reflect from the mirror (Fig. 10.13c). In the case when Bob

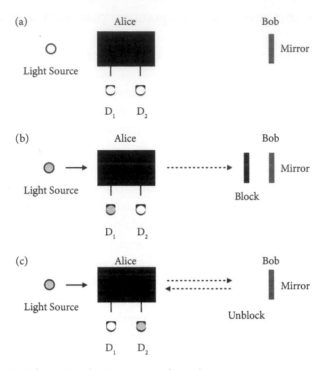

Fig. 10.13 Schematics for the counterfactual communication.

blocks the photon intending to send a "0," the photon is registered at the detector D_1 at Alice's end (Fig. 10.13b). This signals Alice that Bob sent a "0" to Alice. On the other hand, if Bob allows the photon to get reflected intending to send a "1," the photon is registered at detector D_2 at Alice's end (Fig. 10.13c), signaling to Alice that Bob sent a "1." The amazing and highly counterintuitive result is that, in both instances, when the photon is supposed to be bouncing back and forth between Alice and Bob, the probability of finding the photon in the transmission channel is zero. Thus, the photon bounces back and forth from mirrors and other optical instruments placed inside the box at Alice's end and gives a click at D_1 or D_2, sensing what Bob has done (putting his hand in front of his mirror or doing nothing), but without traveling to Bob.

How can this be done? This looks almost magical or supernatural like psychic communication. In order to understand the protocol for counterfactual communication, we first consider some simple interferometric configurations before we explain a system of mirrors and beam splitters that can exhibit counterfactual communication.

First, we consider an array of Mach–Zehnder interferometers, as shown in Figs. 10.14a and 10.14b. In both cases, the reflectivity of the beam splitters in the middle depends on the number of the Mach–Zehnder interferometers and is large when the number of the interferometers is large. In Fig. 10.14a, there is no object on the right side for any of the Mach–Zehnder interferometers whereas, in Fig. 10.14b, there is an object or absorber in the right arm of each interferometer. As before, a single photon is incident on the first beam splitter from the left.

What is the probability of a click at D_1 and D_2 in both cases? It can be shown that, for some appropriate conditions, the detector D_2 clicks with unit probability when there is no object on the right side (Fig. 10.14a) and the detector D_1 clicks with almost unit probability when there is an object at each step (Fig. 10.14b).

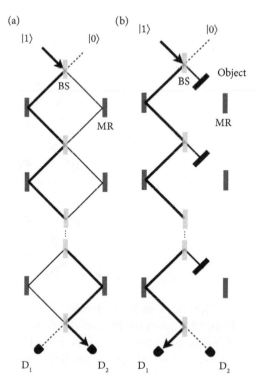

Fig. 10.14 (a) An array of Mach–Zehnder interferometers. (b) Same set-up as in (a) but with an object at each step such that if the photon leaks to the right-hand side, it is scattered and lost.

Next we show that an extension of the interference set-up in Fig. 10.14 can lead to counterfactual communication—communication with no photons present in the transmission channel.

The set-up for counterfactual communication is shown in Fig. 10.15. It also consists of a large array of Mach–Zehnder interferometers with the difference that, instead of single mirrors on the right side, there is an

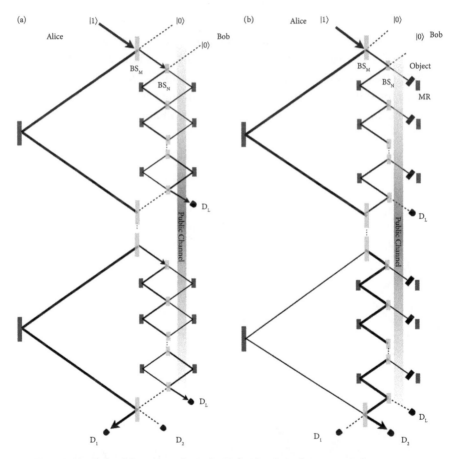

Fig. 10.15 A double array of Mach–Zehnder interferometers for counterfactual communication. Here Alice sends the photon and Bob decides whether to send (a) bit "0" by not blocking the mirrors or (b) bit "1" by absorbing the photon if it enters the Bob side of the interferometer.

array of Mach–Zehnder interferometers of the type shown in Fig. 10.14. A single photon is sent as an input.

The system is set up in such a way that the source of photons, all the beam splitters, all the mirrors on the left side of the outer interferometers, all the mirrors on the left side of the inner interferometers, and the detectors D_1 and D_2 are on Alice's side. The only objects at the Bob's end are the mirrors on the right side of the inner interferometers and the detectors D_L as shown in Figs. 10.15a and 10.15b. The region between Alice's set-up and Bob's mirrors is the transmission region that can be made large.

We now describe the counterfactual communication protocol. In this protocol, Alice sends a single photon from the left as shown in Fig. 10.15, but the information is sent by Bob. He sends a "0" when he does not place any absorbing object in front of all the mirrors as shown in Fig. 10.15a and a "1" by placing an object that absorbs the photon in front of all the mirrors at his end as shown in Fig. 10.15b. This can, for example, be done by placing a long object that covers all the mirrors in Bob's possession on the right side.

We now show that, when Bob sends "0," the detector D_1 at Alice's end clicks and she knows that Bob sent a "0." Similarly, when Bob sends "1," the detector D_2 clicks and Alice knows that Bob sent a "1." The interesting and counterintuitive result is that, in both instances, there is no photon present in the transmission channel.

First, consider the case when Bob wants to send "0" to Alice. He does not block the photon as shown in Fig. 10.15a. For a large number of outer interferometers, there is a large probability that the photon is reflected from the beam splitters BS_M's. However, there is a small probability of transmission. In those rare instances when photon is transmitted from BS_M to the right side, it always ends up at the detector D_L and is lost (see Fig. 10.14a). There is, thus, zero probability that the photon may enter the large interferometer back from right side to left side through the beam splitters BS_M's. Basically the array of the inner Mach–Zehnder interferometers in the right arms of the outer interferometers acts as an absorber. The situation is therefore effectively similar to that of Fig. 10.14b. The result is that, for a sufficiently large number of interferometers, the photon is received at the detector D_1 with an almost unit probability. The important thing to note is that, in those instances when we get a click at D_1, the photon follows an outer trajectory, reflecting from all BS_M from the left side, and is never found in the transmission channel.

Next, we consider the case when Bob wants to send "0" to Alice. He blocks the photon at each stage as shown in Fig. 10.15b. There is an array of blocked Mach–Zehnder interferometers on the right side of the outer interferometers at each step. We have seen, in our discussion of such a blocked array (Fig. 10.14b), that the photon ends up on the left side (at detector D_1) with almost unit probability. Thus, the configuration shown in Fig. 10.15b is effectively the configuration for the outer Mach–Zehnder interferometers (similar to Fig. 10.14a) with no block and the photon is detected at detector D_2 with almost unit probability. A click at D_2 signals Alice that Bob sent a "0." Again, the important observation is that we can be sure that the photon never crossed through the transmission channel, because if it had, it would have been absorbed by one of the blockers.

We thus achieved a highly counterintuitive result: When Alice sends a photon, Bob can send a "1" by allowing the photon to be reflected from his mirrors and Alice gets a click at detector D_1 AND Bob can send a "0" by blocking the photon at each stage and Alice gets a click at D_2. In both instances, the probability that the photon exists in the transmission channel is zero.

This work has been described as startling, counterintuitive, and mind-boggling. For example, Yakir Aharonov and Daniel Rohrlich, in a *Physical Review Letters* article, expressed their astonishment in these words:

> H. Salih, Z.-H. Li, M. Alamri, and M. S. Zubairy [1] describe a remarkable effect, which they call "counterfactual quantum communication": transmission (across a "transmission channel") of information from a sender to a receiver "without any physical particles traveling between them." ... For all our familiarity with quantum nonlocality, the effect is startling... If any effect evokes Einstein's famous phrase "spooky action at a distance," it is this one.

In a popular article with the title *The quantum carrier pigeon that wasn't there* published in *Physics Today*, S. Slussarenko, N. Tischler, and G. F. Pryde describe this work as follows:

A scheme for true counterfactual communication was proposed in 2013 at Texas A&M University by Muhammad Suhail Zubairy and coworkers. ... In their recent work, Pan and colleagues experimentally verified Zubairy team's scheme ... It's difficult to grasp the implication of this experiment that seemingly sent data from Bob to Alice without any physical messenger passing between them.

11

Schrödinger's cat and entanglement

If quantum mechanics hasn't profoundly shocked you, you haven't understood it yet.

— *Niels Bohr*

Wave–particle duality lies at the foundation of the conceptual foundation of quantum mechanics. As discussed in the last chapter, this counterintuitive concept is highlighted via Young's double-slit experiment. For example, electrons are known to act as particles in most experiments. However, they demonstrate wave nature by forming interference fringes in the double-slit experiment. We also noticed that this is a very mysterious behavior. The electrons are sent one by one via a double slit and detected on a screen farther away. Each electron is detected on the screen like any other particle, such as a bullet. However, when a large number of electrons are detected on the screen, they show an interference pattern. This is mind-boggling—how could the electrons know where to hit the screen such that the resulting pattern is similar to the pattern that we associate with a light wave.

We noted that this amazing behavior could only be explained via a wavefunction approach. Each incident electron was treated as a wave packet that passed through the two slits simultaneously. Only such a description can successfully explain the appearance of the interference fringes on the screen. If each electron is assumed to be passing through one slit or the other, it would be impossible to explain the interference pattern.

The simultaneous passage of the electron through the two slits can be treated as the electron being in a coherent superposition of states. Here the states correspond to being at the two individual slits. This description does not imply that each electron is smeared over the two slits, with half the electron passing through one slit and the rest through the other slit. What this wavefunction approach implies is that we have no knowledge whether the electron is at one slit or the other. Only when we try to find out which slit the electron passed through do we know which one it did. However, in that case the interference disappears.

The coherent superposition of states is a much wider phenomenon and is among the most mysterious aspects of quantum mechanics. Even several founders of quantum mechanics, most notably Erwin Schrödinger, felt very uncomfortable with this notion. A related phenomenon involving not just the individual particle but two or more particles is quantum entanglement where the particles influence each other no matter how far away they are from each other. Two particles can be, in principle, in different parts of a galaxy, but what is done to one particle can instantaneously affect the other.

In this chapter, the concepts of coherent superposition and entanglement are elucidated. But this is not all. These counterintuitive behaviors are no longer a matter of intellectual curiosity. They provide the applications in important areas such as quantum teleportation and quantum computing that we discuss in the later part of this chapter.

11.1 Coherent superposition of states

If we have a system that can exist in two possible states then it is found either in one state or another but never simultaneously in both states. For example, a door can be either open or it can be shut but cannot be in a state where it is both open and shut. Similarly, a ball can be inside a box or outside the box but never be simultaneously in "inside" and "outside" states.

A quantum system, on the other hand, can be in a "coherent superposition of states."

In order to grasp this concept, we revert to the example of the fictitious quantum needle (Section 6.1) and then the realistic example of photon polarization.

We recall that, when the quantum needle in a vertical state passes through a screen oriented at 45 degrees, there is a 50 percent probability for passing through the door oriented at 45 degrees and a 50 percent probability that it will pass through the door oriented at 135 degrees.

Another way of looking at this situation is as follows. First, we interpret the needle in vertical state to be in an equal superposition of two states: a state oriented along 45 degrees and a state oriented at 135 degrees. This superposition state persists as long as we do not measure the orientation of the needle. However, when we measure the state of the needle by passing through the screen oriented at 45 degrees, the state of the needle "*collapses*" and we find it either oriented at 45 degrees or 135 degrees.

At first sight this result does not appear surprising or mysterious. After all we seem to encounter similar situations all the time. If a window remains open fifty percent of the time and remains closed for the remaining fifty percent time then, before looking at the window, we can only say that there is a 50 percent probability that it is closed and 50 percent probability that it is open. However, when we look at the window, it is found either open or closed—never simultaneously open and closed. So, what is so special about the quantum "coherent superposition of states"?

The mysterious behavior is that, if, for the same quantum needle in the vertical state, the screen is rotated by any other angle, it will be in a superposition of states of mutually perpendicular orientations. This behavior is quite different from our everyday objects like a door. For example, if the screen is rotated by (say) 30 degrees, the needle can be interpreted in a coherent superposition of the states with orientations 30 degrees and 120 degrees. The corresponding probabilities are 1/4 for passage through the door with 30 degrees orientation and 3/4 for the 120 degrees orientation. What this means is that, if a large number of identical needles (say 1000) pass through the screen, roughly 250 will be found along 30 degrees and 750 along 120 degrees.

The same behavior is expected for photons with polarization: A photon in state ↑ can be considered to be in an equal superposition of states ↗ and ↖. When this photon passes through a polarizing beam splitter rotated by 45 degrees, there is an equal probability of finding the photon in state ↗ and in state ↖. Also, if the same photon (in state ↑) passes through a beam splitter rotated by 30 degrees, the photon will be interpreted in a linear superposition of two state of polarization, states with orientations along 30 and 150 degrees. The passage probabilities will again be equal to 1/4 along 30 degrees and 3/4 along 120 degrees.

This behavior is very different from any real-world object. The quantum state has no objectivity, the state depends on the setting of the measurement apparatus. Before making a measurement, it would appear that the object exists simultaneously in both states. The measurement results in the *collapse* to one of the two states that are available through the setting of the measurement apparatus. This explains the counterintuitive nature of the quantum property called "coherent superposition of states."

11.2 Schrödinger's cat paradox

Erwin Schrödinger was one of the founders of quantum mechanics. However, he always felt uncomfortable with quantum mechanics and came up with paradoxical consequences. Schrödinger's cat paradox (1935) is one such example. Schrödinger's cat experiment is an example of a *gedanken experiment*, i.e., we do not *actually* conduct the experiment, we use only our imagination and reasoning instead.

What Schrödinger was concerned with was the notion of the coherent superposition of states where, as we have seen above, the system can be simultaneously in two quantum states. However, when we measure, we find the system in one state or another. Schrödinger argued that if we extend the ideas that are valid for microscopic objects such as an electron, a photon, or an atom to the macroscopic world, it may lead to absolutely amazing consequences. One such consequence is a cat that can simultaneously be in a state of being alive and a state of being dead. Since we do not encounter a cat simultaneously both alive and dead, there must be something wrong about quantum mechanical interpretation.

In Schrödinger's *gedanken* experiment, a cat is placed inside a steel box. The box also contains a radioactive atom, a Geiger counter, a hammer, and a vial of poison. The radioactive atom can decay in the process by which its unstable atomic nucleus loses energy by emitting massive alpha and beta particles and massless gamma ray photons. Radioactive decay is a random process at the level of single atoms. According to quantum theory it is impossible to predict when a particular radioactive atom will decay. A Geiger counter can detect the emitted particles. In Schrödinger's set-up, when the radioactive substance decays, the Geiger counter detects it and triggers the hammer to release the poison, which subsequently kills the cat (Fig. 11.1).

Thus, we have two possibilities: the radioactive decay does not take place and the cat is alive OR the radioactive decay does take place and the cat is dead. We thus have the microscopic superposition of the state of the radioactive atom leading to a state for the cat that is in a macroscopic superposition of state of being alive and dead at the same time. This is a paradoxical situation, as in real life, we do not see a cat that is simultaneously alive and dead.

Until the box is opened, an observer does not know whether the cat is alive or dead. The cat's fate is intimately tied to the atomic state. If the atom has not decayed, the cat is alive but if it has decayed, the cat is dead. As soon as the box is opened, we find that the cat is either dead or alive but not in a "superposition" state—cat is alive or cat is dead but not both.

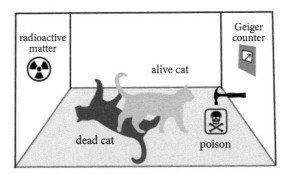

Fig. 11.1 Schrödinger's *gedanken* experiment: A radioactive atom decays and the radiated particle is detected by a Geiger counter that releases a hammer hitting a vial of poison, causing the cat to die. So the state of the cat (dead or alive) depends on the state of the radioactive atom (dacayed or not decayed).

Here we present the cat paradox in Schrödinger's own words:

One can even set up quite ridiculous cases. A cat is penned up in a steel chamber, along with the following device (which must be secured against direct interference by the cat): in a Geiger counter, there is a tiny bit of radioactive substance, so small, that perhaps in the course of the hour one of the atoms decays, but also, with equal probability, perhaps none; if it happens, the counter tube discharges and through a relay releases a hammer that shatters a small flask of hydrocyanic acid. If one has left this entire system to itself for an hour, one would say that the cat still lives if meanwhile no atom has decayed. The first atomic decay would have poisoned it. The psi-function of the entire system would express this by having in it the living and dead cat (pardon the expression) mixed or smeared out in equal parts ... It is typical of these cases that an indeterminacy originally restricted to the atomic domain becomes transformed into macroscopic indeterminacy, which can then be resolved by direct observation. That prevents us from so naively accepting as valid a "blurred model" for representing reality. In itself, it would not embody anything unclear or contradictory. There is a difference between a shaky or out-of-focus photograph and a snapshot of clouds and fog banks.

An interesting and revealing comment is by Einstein in his letter to Schrödinger, essentially patting him on the back, for coming up with an elegant argument regarding "reality"[1]:

You are the only contemporary physicist, besides Laue, who sees that one cannot get around the assumption of reality, if only one is honest. Most of them simply do not see what sort of risky game they are playing with reality—reality as something independent of what is experimentally established. Their interpretation is, however, refuted most elegantly by your system of radioactive atom + amplifier + charge of gun powder + cat in a box, in which the ψ-function of the system

---

[1] Einstein's comment on Schrödinger's cat in a letter to Schrödinger dated December 22, 1950 is in K. Przibram (ed), Letters on Wave Mechanics (Philosophical Library, New York, 1967), p. 39.

contains both the cat alive and blown to bits. Nobody really doubts that the presence or absence of the cat is something independent of the act of observation.

We discuss Einstein's concept of reality and his own criticism of quantum mechanics in detail in Chapter 12.

11.3 Quantum entanglement

The ability of a quantum system to exist in a coherent superposition of states is a novel feature of quantum mechanics. Another interesting aspect of quantum systems is that they can exist in an entangled state. Quantum entanglement is not only a counterintuitive effect but, as we see in our discussions on quantum computing in Section 11.5, it is a remarkable resource.

Let us consider a system of two independent objects. In principle, they may be so far apart that they cannot interact with each other in any way. If this happens then, according to classical mechanics, both objects are independent of each other and their properties are not influenced by what we do to the other object. For example, if we have two balls, one of them is red and the other is blue. Let them be very far away from each other. No matter what we do to the ball in our possession (paint it, crush it, throw it ...) the properties of the ball far away from us will not be affected in any way. This behavior can be seen in all classical systems.

Quantum mechanical systems can remarkably demonstrate a dramatically different behavior. We illustrate it with a simple example. Let us consider two photons A and B. Suppose we prepare them in the quantum state such that they both have the same direction of polarization. Let us also suppose that they move in opposite directions and are very far away from each other. If photon A is horizontally polarized then photon B is also horizontally polarized and if photon A is vertically polarized then photon B is also vertically polarized. Thus, if Alice measures the A photon via a polarizing beam splitter, and if she finds the photon to be horizontally polarized then she can be certain that if Bob measures the B photon with the same setting of the beam splitter, he will also find his photon to be horizontally polarized. So far there is no mystery. The situation is completely analogous to the following example.

Let there be a box with two balls inside. Alice and Bob are told that both balls are either red or blue. If Alice and Bob pick one ball each and, without looking, move in opposite directions. When they are too far apart to communicate, Alice looks at the ball in her possession. If it is found to be red then she instantaneously concludes that the ball in Bob's possession must be red as well. The quantum situation is, however, more complex.

Let Alice and Bob both rotate their polarizing beam splitter by (say) 30 degrees. Then, if Alice finds the A photon polarized at 30 degrees, Bob will also find the B photon polarized at 30 degrees. Similarly, if Alice finds A photon to be polarized along 120 degrees, Bob will also find photon B to be polarized along 120 degrees. Indeed, this correlation persists for any angle of rotation of the polarization beam splitter. This is a highly counterintuitive result. How come photon B adjusts its state of polarization depending on the setting of the polarization beam splitter very far away. The state of Bob's photon depends on what Alice decides to do even when there is no way that Alice's and Bob's photons can interact with each other. The same is also true for Alice's photon—the state of the photon in Alice's possession is influenced by what Bob does to his photon. The two photons are "*entangled*" even if they are far apart.

There is no corresponding result for classical objects. No classical system can demonstrate this amazing behavior. The two photons A and B are said to be in an "entangled state."

There is another way that quantum entanglement between two objects displays a behavior that is not possible for any conventional or classical system. For a classical system like two balls, one red and the other blue, if we know everything about the two balls (their color, shape, size, etc.) then we can claim to know everything about the individual ball as well. The same cannot be true for a quantum system.

As an example, we consider two photons which are in an entangled state. If one is found with horizontal polarization, the other is also found with horizontal polarization. Similarly, if one is found in the state with vertical polarization, then the other is also found in the vertical polarization state. Thus, we have complete knowledge about the "two-photon system." However, we do not have any definite prediction for an individual photon. Both the photons can have either horizontal polarization or vertical polarization. But we cannot predict the polarization direction of either photon. Our knowledge for the individual photon is fuzzy.

This property of two or more objects where we have complete knowledge of two or multiple objects but only a fuzzy knowledge of the individual particles is a quantum property with no classical analog.

11.4 Quantum teleportation

Teleportation is a science fiction concept that was popularized by the 1960s TV program "Star Trek." The phrase "Beam me up, Scotty!!" used by Captain Kirk to direct his Chief Engineer Scott when he wanted to be transported from a faraway planet to the Starship Enterprise, remains one of the most famous phrases of television programming. Teleportation was used as a means of transport—Captain Kirk would disappear on the planet and reappear on a platform inside the starship.

Can we do teleportation in real life? If teleportation becomes possible, there will be no need for cars or air planes—we could be teleported from one place to another. So far, teleportation of human beings is only science fiction—it would take an infinite amount of energy to teleport even a single atom. However, teleporting a two-state quantum system has become a reality, thanks to the concept of quantum entanglement. Quantum teleportation of a two-state system from one location to another, first proposed by C. H. Bennett, G. Brassard, C. Crépeau, R. Jozsa, A. Peres, and W. K. Wootters, is one of the most beautiful examples of quantum entanglement.

The problem of teleportation of a quantum state can be described as follows. Suppose Alice has a photon polarized in an unknown direction at location A. Suppose there is another photon located at Bob's end C. Just like in $Star$ $Trek$, we would like to destroy the unknown quantum state of the photon at Alice's end and create the same quantum state of the photon at Bob's end.

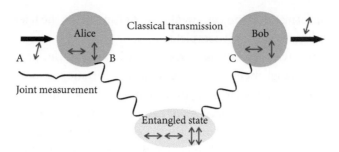

Fig. 11.2 Schematics for quantum teleportation.

The quantum teleportation of the state A at Alice's end to Bob is accomplished in three steps.

In the first step, two photons B and C are prepared in an entangled state as shown in Fig. 11.2. Photon B is sent to Alice and photon C is sent to Bob. This is done via what we call a quantum channel. At the end of this step, Alice has both the photons A (whose unknown state has to be teleported) and the photon B (one half of the entangled state).

The second step is the most crucial. It uses the characteristics of entanglement as discussed in previous sections. When two systems are entangled, if one system is disturbed, the state of the other is also affected no matter how far away it is. The key to quantum teleportation is that Alice makes a joint measurement of the two photons in her possession in such a way that the quantum state of the photon C in Bob's possession changes to state of Alice's photon A. However, another feature of quantum mechanics prohibits us from getting a unique outcome in this step. There are four possible measurement outcomes of the two photons A and B, each with a probability of 25 percent. One of these outcomes transfers the quantum state A to Bob's photon C. If this happens then the teleportation of photon A to state C succeeds. However, there are three other possible outcomes. Each of these outcomes transfers the quantum state A to C in such a way that a simple manipulation of the C photon can be restored to state A.

However, at the end of the second step, Bob does not know the outcome of Alice's measurement. It can be any one of the four possible outcomes. In the third and the final step, Alice informs, through a classical channel like a telephone, the outcome of her measurement. Based on this information, Bob can tweak his photon so that the teleportation of the unknown state A is complete.

Quantum teleportation is a beautiful example of the properties of quantum entanglement. The process of teleportation is not instantaneous, because, in the last step, information must be communicated via conventional means between Alice and Bob as part of the process. The usefulness of quantum teleportation lies in its ability to send quantum information arbitrarily far distances without exposing quantum states to thermal decoherence from the environment or other adverse effects. It has the potential of being useful in communication technology and may be helpful in developing a "quantum internet."

Another area where quantum entanglement plays a crucial role is quantum computing. This topic is discussed in the next section.

11.5 Quantum computing

The emergence of quantum computing as a major field of research is due to the realization that certain problems can be solved much faster on a quantum computer than on a conventional computer. The extraordinary speed of a quantum computer is due to the novel features of quantum mechanics, such as coherent superposition and quantum entanglement as discussed above. The computational power of a quantum computer can exceed that of conventional computers.

The first ideas about quantum computing can be traced to Richard Feynman when, in 1982, he proposed solving complicated quantum mechanical problems by simulating them on a quantum computer. The mathematical framework of quantum computing was developed by David Deutsch in the late 1980s. There was, however, no practical problem that could be solved on a quantum computer substantially faster compared to conventional computers. The interest in the field of quantum computing remained limited to a very small number of researchers. The situation changed drastically in the mid 1990s when two major quantum computing algorithms were proposed. One related to factoring a large number into its prime factors and the second related to finding a marked object in an unsorted database. Since then, there has been tremendous activity in this field. Quantum computing provides a beautiful example of how quantum mechanical concepts such as coherent superposition of states and quantum entanglement can lead to incredibly fast speed-up in the solution of certain problems.

The basic building block of a computer is a bit that can take on two values, "0" or "1." In a conventional computer, these bits are classical objects like voltage, high voltage corresponds to "1" and low voltage corresponds to "0." These are therefore referred as classical bits or simply as "bits."

On the other hand, a quantum bit (or a "qubit") is a system that can exist in two possible quantum states. The qubits, in the laboratory, can correspond to many different realizations. For example, a photon that can exist in the polarization state → or ↑ corresponding to "0" or "1," respectively, is a qubit. Other examples include an atom in the ground state G and the excited state E and a radiation field in states with no photon or one photon corresponding to "0" or "1," respectively. Any computer that carries out computation using qubits, instead of classical bits, is called a quantum computer.

Quantum bits, being quantum states, can satisfy two novel properties which are inaccessible to their classical counterparts. The first is the possibility that qubits can exist in a state of coherent superposition. For example, an atom can exist in the ground state G or in excited state E or a coherent superposition of states G and E. Secondly, a system consisting of two or more qubits can be found in an entangled state where the two qubits lose their independent identity and the state of one qubit depends on the state of the other as discussed in the previous section.

In order to appreciate the potential power of the quantum computer, let us first consider a simple classical example. Suppose there is one window. It can be in two states: open and closed. If there are two windows, there are four possibilities: both are open, the first is open and the second is closed, the first is closed and the second is open, or both are closed. For three windows, there will be $2 \times 2 \times 2 = 8$ possibilities. If there are 256 windows, the number of possibilities are $2 \times 2 \times \cdots \times 2$, 256 times ($2^{256}$ in scientific notation). This is a huge number, larger than all the atoms in the entire universe. However, in spite of the large number of possibilities, there is only one possibility that exists. For example, in the case where we have only two windows, we will find one of the possibilities. Either both are open, the first is open and the second is closed, and so on.

The situation is markedly different for quantum mechanical objects. Consider a quantum system, like an atom, that can exist in two possible states, ground and excited. If there are 256 atoms, then, in principle, they can exist in all the possible combination of states simultaneously—in a massive entangled state. Thus, if we have the capability to manipulate just 256 qubits, such as 256 atoms, we can have a computer with a power that can be acquired by our conventional computers only if they work with all the atoms in the known universe as bits. This realization is scintillating, almost mind-boggling. This observation also clearly shows why one feels extremely excited at the prospect of a quantum computer.

Then comes the obvious question: If, even with the modest number of 256 qubits, we can potentially make a computer that will match a conventional computer with all the atoms in the known universe in its memory, why is such a computer not a reality so far? The answer to this question lies in two aspects of quantum mechanics that would appear to destroy the prospect of a quantum computer. Just as the quantum mechanical concepts of coherent superposition and quantum entanglement lead us to the prospect of a quantum computer with almost unbelievable capability, the other two aspects of quantum mechanics, namely the probabilistic nature of the measurement outcome and decoherence, can prove fatal to the design of a robust and reliable computer.

The probabilistic nature of quantum mechanics has been emphasized in Chapter 7. In the present context, the entangled state of the 256 qubits exists until we measure the state of the qubits. However, if a measurement is carried out on each qubit, the outcome is probabilistic. Out of 2^{256} ($2 \times 2 \times 2 \cdots \times 2$, 256 times) possibilities, only one possibility will be realized probabilistically. This is very disappointing as a probabilistic answer to any computation is typically not acceptable. So, this should be the end of the "quantum computer" fantasy.

The second issue relates to decoherence. It is more subtle. It has to do with the errors induced in the state of qubits by their interaction with the environment. In our everyday experience, we know that any interaction of a given system with the environment can lead to a change in the state of the system even when we do not intend to do so. For example, a cup of hot coffee can cool to room temperature by losing energy to the environment. The environment is huge and we do not notice any change in the temperature of the environment. However, the temperature of the coffee in our little cup goes down. This process is irreversible as the cooling of the coffee can and does take place by losing energy to the surroundings but it never happens that a cup of cold coffee suddenly heats up by extracting energy from the surroundings. Quantum mechanically the situation is even more complex. It has to do with the nature of vacuum. As discussed in Chapter 9, before the advent of quantum mechanics, vacuum was perceived as *nothing*—a place where no light existed, nothing moved, and there was no energy present. The quantum mechanical picture of vacuum turned out to be dramatically different. According to quantum mechanics, vacuum, even at almost absolute zero temperature when nothing moves and no energy is supposed to be present, has an infinite amount of energy associated with an electromagnetic field. In addition, there are quantum mechanical fluctuations as a result of Heisenberg's uncertainty relation that cannot be neglected. Thus, when a qubit, such as an atom in the excited state, experiences fluctuating fields associated with the vacuum, it can decay to the ground state spontaneously. This becomes a source of inevitable error.

Thus, we have a situation where, on the one hand, the possibility of preparing a qubit in a coherent superposition of its states and the possibility of multiple qubits to exist in an entangled state raises the possibility of processing a large (indeed extremely large) amount of data much faster, but, on the other hand, the probabilistic outcome of the measurement and the presence of inevitable decoherence due to the interaction of the qubits with the environment appear to kill the prospect of quantum computing. The challenge, therefore, is to come up with problems that can be solved by taking advantage of the tremendous potential of quantum entanglement but are not adversely affected by the probabilistic nature of measurement outcome. Decoherence is still a stumbling block. This is a practical problem that requires searching for the systems where we can minimize the effect of decoherence or using some error correction codes and procedures. This challenge is met in a limited number of problems—the two most famous being the determination of the prime factors of a composite number (the so-called Shor's algorithm) and the search of an unsorted database (the so-called Grover's algorithm).

11.6 Shor's algorithm

Prime numbers are numbers that are not divisible by any number other than themselves and 1. Examples are numbers such as 7, 13, and 23. Anyone with a pocket calculator knows it is trivial to multiply two prime numbers, however large they may be. For example, 113 and 169 are prime numbers. They can be multiplied easily and the result is 19,097. However, carrying out the reverse—finding the unknown prime factors given a large number—is extremely difficult. To appreciate the difficulty, try to find the prime factors of the number 21583^2.

In order to appreciate the difficulty of factorizing a large number, we note that it would take decades to find prime factors of a 256-decimal digit number on one of the fastest computers available today. A 1000-decimal digit number would take about ten billion years. This is indeed a long time considering that the present estimate of the age of the universe is 13.8 billion years. Thus, by increasing the size of the numbers, the amount of difficulty in finding the prime factors of the number is made extremely large. In 1994, Peter Shor proposed a scheme based on a quantum computer to find the prime factors of a large number very efficiently.

Finding the prime factors appears to be a rather mundane problem of pure mathematics with no applications. This is not true—the present-day multi-trillion-dollar e-commerce would collapse if a scheme can be found to factorize large numbers in a small time.

As discussed in Section 6.3, conventional cryptography was used for secure communication for millennia until around 1970, when the advent of public key algorithms changed the nature of cryptography dramatically. Even before the internet became a widely used means of communication, there was a need for a key exchange through a public

---

[2] The answer can be found in the footnote on the last page of this chapter.

channel. If a bank wants to communicate sensitive financial information with branches scattered all over the world, it would be too impractical and expensive to exchange keys via physical contact. The key has to be exchanged on communication channels that are readily available to everyone, such as a telephone line or an internet. With the arrival of the internet in the 1990s and its usefulness in e-commerce, it became essential that data and information transfer should be exclusively on public channels, and, at the same time, should be secure.

Three mathematicians, Ronald Rivest, Adi Shamir, and Leonard Adleman, devised the so-called RSA protocol in 1978 that allows an exchange of keys through public channels with a very high level of security. The RSA protocol is a beautiful (and useful) example of the purest branch of mathematics, called number theory.

The RSA public key system is schematically described in Fig. 11.3. Here Alice wants to exchange information with many friends Bob1, Bob2, ... scattered around the world in complete security. She would like to do it in such a way that if Bob1 sends some information to Alice, then Bob2, Bob 3 etc. should not be able to figure out the message even if they have access to the information that Bob1 sent to Alice.

The way this is done is that Alice chooses a large number that is a product of two prime numbers p and q, $N = pq$. From these three numbers N, p, and q, Alice generates two more numbers e and d using certain methods of number theory. Then the encryption key consists of two numbers e and N that Alice announces to all her friends on some public channel like the internet, a cell phone, or a telephone line. This

Fig. 11.3 Schematics of RSA key distribution. Alice announces the encryption key (e, N) to everyone including her friends as well as potential eavesdroppers. She keeps the decryption key d secret.

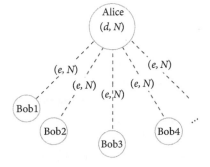

encryption key is accessible to any eavesdropper as well. Alice retains the decryption key d that is kept secret. Now if any of Alice's friends wants to send a message (a decimal number) he/she can encode the message with the key e and N in a prescribed manner and send to Alice. Alice, at her end, uses the decoding key d to decipher the message. The amazing aspect of the RSA protocol is that, although all the friends and eaves-droppers have access to both the encoded message and the encoding key (e and N), it is almost impossible for them to decipher the message. This seems almost too good to be true—announcing the key to every-one with no one being able to decipher the sent message. How is this possible?

To see this, we note that what is public knowledge in the RSA protocol are the encryption key e and N. In order to break the security, one needs to know the decryption key d and that, in turn, requires a knowledge of the prime numbers p and q. Thus, the difficulty of factorizing N ensures the security of the RSA protocol.

In this background, it is easy to understand that, when, in 1994, Peter Shor proposed the quantum computing algorithm to find the prime fac-tors p and q of the number N much more efficiently, it sent a major alarm throughout the international community. The potential of quan-tum mechanics to seriously attack the foundation of e-commerce sent tremors through the security community and created a great interest in the emerging field of quantum computing. Shor's algorithm remains the most powerful example of the applications of quantum computing.

11.7 Quantum shell game

Another important application of a quantum computing algorithm is to find an object in an unsorted database. Before discussing this impor-tant application of quantum computing, we discuss a shell game that we may have played in our childhood. We show how a quantum com-puter can help in playing this game and winning each time in an almost unbelievable way.

Consider four inverted cups or nutshells with only one of the shells hiding a pea underneath. After shuffling them, the contestants must spot

Fig. 11.4 In the shell game, contestants must spot the shell with a pea underneath.

the shell with the pea underneath (Fig. 11.4). The search process requires inverting each shell one by one to find the pea. If we are lucky, we may find the pea under the first shell and win. The probability of this happening is only 25 percent. If we are not lucky, we may have to flip all four before we find the shell with the pea. On the average it requires more than two searches to find the pea.

Can we find the pea in only one try with certainty every time? Conventional wisdom tells us that this is simply not possible. However, the situation is different in the quantum shell game. In a shell game, where the shell is replaced by a quantum state and the pea is replaced by an "inverted" target state, we can find the target state with certainty with only one measurement.

We now discuss Grover's algorithm, for the search of an object in an unsorted database. The searching problem can be stated as follows: Suppose there is a database consisting of N items out of which just one item satisfies a given condition. We would like to retrieve that item.

There are two kinds of databases, a sorted database and an unsorted database. Consider a telephone book with names listed in alphabetical order with their telephone numbers. If we are given a name and want to find the telephone number, we can search alphabetically to recover the telephone number. This is an example of a sorted database. However, what if we are given the telephone number and we would like to know the name of the person it belongs to? This is an example of an unsorted database as the telephone numbers in the telephone book are listed completely randomly. In the search of an item in an unsorted database of N items, we would look at each item one by one. If we are lucky, the first item can be the searched item. However, there is a possibility that the searched item is the last on our search list, thus requiring N searches. On the average, it may require $N/2$ searches before we find the desired item. In the example of the telephone book, if we are searching the name of the person with a certain telephone number, we may require, on the average, a search through half the telephone book before we see the name matching the telephone number. This is the essence of what we call the classical search.

Next, we ask the question: Can quantum mechanics provide a speed-up? Can we do better than $N/2$? It has been shown that, for a certain class of search problems, the searching process of an unsorted database in a quantum computer may require only $\sqrt{N}$ searches instead of $N/2$ for large databases.[3] Thus, for example, half a million searches may be required to search a marked item in a database of one million items, whereas only 1000 searches are required on a quantum computer.

---

[3] The prime factors of 21583 are 113 and 191.

12

Is reality really real?

We often discussed his (Einstein's) notions on objective reality. I recall that during one walk Einstein suddenly stopped, turned to me and asked whether I really believed that the moon exists only when I look at it.

– Abraham Pais

The question of the *reality* of objects, properties, and phenomena has been addressed since ancient Greek times. What is reality? Perhaps, the most symbolic of the discussion on reality is the *Allegory of the Cave* by Plato in his book *Republik*. With a deep understanding of the laws of nature, what is the status of our understanding of reality in the twenty-first century?

An object can be considered to be *real* if it exists independent of whether we look at it or we do not. For example, we all believe that the moon exists even when none of us looks at it. It is hard to imagine that the reality of an object like the moon depends on us directly observing it. We cannot imagine a world where objects exist only when we look at them and cease to exist when we turn away from them. At night, we can be comfortable in our thoughts that the sun is there even when none of us is able to look at it.

As we see in this chapter, quantum mechanics seems to deny the existence of an objective reality, a reality that exists independent of the presence of observers. This is a stunning result if it is true.

In the previous chapters, we have repeatedly seen that measurement is crucial in studying the properties of a system. For example, as discussed in Section 5.5, the electron in a hydrogen atom is described by a wave-function whose square yields the probability of finding the electron at a certain location. This is all that quantum mechanics tells us. When we try to find the electron, it is found at a certain location. The probability of finding the electron is higher where the value of the square of the wavefunction is higher. This appears to be a rather unsatisfactory situation. Our conventional concept of science is that it should be able to make definite predictions. If a theory is able to give only the probability of an event (it will happen or it will not), it should not be a useful theory. Some, including Albert Einstein, came to believe that quantum mechanics with its probabilistic predictions cannot be a complete theory and a search should continue for a "complete theory."

But the problem is even deeper.

A fundamental question is whether the electron existed at the location where we found it *before* the measurement was made. Does the probabilistic description of quantum mechanics represent only our lack of knowledge or is it something more fundamental? Can we say that, before our measurement, the electron was located at that position and we just did not know? Or the electron had no real position or even existence before our measurement and it acquired the particular position as a consequence of the measurement? These represent the question of *reality* that is addressed in this chapter. The answer to these questions is nothing what our classical intuition tells us.

Einstein supported the view that physical objects like electrons and atoms exist as real things and they exist independent of an observer. After all we believe that the universe existed for a long time before the first life and first human appeared on earth. However, this view, that appears obvious and quite true, is fatally flawed according to the laws of quantum mechanics.

The point of view that prevailed is due to Niels Bohr and Werner Heisenberg. According to them, nothing, not even electrons and atoms, can be proven to be real until they are looked at. To them any object or property is just an idea until we ask questions about nature and then nature supplies the answer to the question asked. We elaborate this statement later in this chapter with the example of the polarized photon and the polarizing beam splitter.

The crucial question is: What happens to the objects or properties that lie outside our questions? Quantum mechanics claims that it is not even meaningful to talk about these objects and properties. They are not real, just ideas.

12.1 Reality and locality

One of the most dazzling and astonishing results of the twentieth century is the realization that *reality* and *locality* do not coexist. This amazing discovery lies at the boundary between science and philosophy.

Like reality, locality is also a deeply rooted concept in us. *Locality* means that whatever happens at a particular location at a given time is not influenced in any way by what happens far away at the same moment. In the twentieth century, a much more formal definition has emerged as a result of Einstein's theory of relativity (Chapter 15). A founding principle of the theory of relativity is that no information can be sent faster than the speed of light in vacuum. This means that if two objects are separated by a distance that light takes one hour to go from one to the other then what happens to one object (destroyed, divided, rotated etc.) cannot be influenced at all by the other object in any way whatsoever for one hour.

Reality and locality have the status of self-evident truths. Therefore, the observation that, according to the laws of quantum mechanics, the concepts of *reality* and *locality* cannot be simultaneously true, and experiments agree with the predictions of quantum mechanics, becomes a truly mind-boggling result. This means that, at least one of our cherished "truths" is not true. If there has to be a reality, it should be non-local.

In our own times, Albert Einstein presented the concept of reality which, in his words, is stated as follows:

> If without in any way disturbing a system, we can predict with certainty (i.e., with probability equal to unity) the value of a physical quantity, then there is an element of physical reality corresponding to this physical quantity.

First, we address the question: How do we establish the reality of an object "without in any way disturbing" or making measurement on it?

How do we know that the moon exists even when none of us looks at it?

A way to establish reality, "physical reality whose existence is independent of human observer," is to look at events with common cause. For example, in order to establish the reality of the moon, we can go to the shore of an ocean and look at the tides. These tides are the result of the gravitational force exerted on the ocean waves by the moon. A careful scientific analysis of the tides (how often they happen and how high they are etc.) can help in proving the existence of the moon without ever looking at it.

As another example we consider a box containing two balls, one blue and the other red. Let two persons, Alice and Bob, blindfolded pick one ball each and travel in opposite directions to distances so far away that they cannot influence each other in any way. Alice then removes her blindfold and discovers that the color of the ball in her possession is red. She instantaneously concludes that the color of the ball in Bob's possession must be blue. She has arrived at this conclusion "without in any way disturbing" the ball in Bob's possession. Therefore, according to Einstein's definition, there is an element of reality associated with the blue color of the ball in Bob's possession.

These simple experiments do not have any mystery about them. However, a version of a similar experiment done on quantum objects like atoms or photons leads to paradoxes. Quantum mechanics does not agree with our common-sense analysis and predictions. Before analyzing the experiment, let us consider another set-up that should explicitly show how quantum mechanics is different from our everyday experience.

12.2 Einstein–Podolsky–Rosen paradox

It is ironical that both, Albert Einstein and Niels Bohr, played crucial roles in laying the foundation of quantum mechanics but strongly disagreed on the interpretation and limitations of the theory. In Chapter 10, we discussed the first round of the debate between Einstein and Bohr when Einstein challenged Bohr's principle of complementarity at the

Solvay conference in 1927 and Bohr successfully defended it. The most serious challenge, however, came in 1935 through a paper by Einstein, Podolsky, and Rosen (EPR) entitled "Can Quantum-Mechanical Description of Physical Reality Be Considered Complete?" In this paper, EPR argued about the incompleteness of quantum mechanics through a *gedanken* experiment.

According to EPR, if a physical quantity is real according to the definition given above, then the theory should be able to predict the value of that quantity precisely. If it is unable to do so, then the theory should be regarded as incomplete.

In a simplified version of Einstein's analysis, let two particles move in opposite directions in such a way that their center of mass remains constant. This means that if two objects of equal mass are moving in opposite directions then the distance from the center will be the same for both objects. This situation can be envisaged as a massive ball, initially at rest, exploding in two equal pieces is such a way that they move in opposite directions with equal distance from the initial position and equal speed at all times. In other words, the two particles are highly correlated. Therefore, by measuring the position of one particle, we can infer the precise position of the second particle even when this particle is so far away that its position is not affected in any way by the first particle. Similarly, in another experiment, if we measure how fast the first particle is moving, we can determine precisely how fast the second particle is moving again without disturbing it in any way.

Therefore, both the position and the speed of the second particle can be determined precisely without disturbing the particle itself. Einstein argued that this means the position and the speed of the second particle are *real* properties.

We therefore have a situation such that both the position and the speed of the second particle have well-defined values according to EPR and an element of reality should be associated with these values. However, as seen in Chapter 8, quantum mechanics prohibits us from assigning precise values to these properties. According to the Heisenberg uncertainty relation, one cannot assign precise values to the position and the speed of a particle at a given time. There is, therefore, a contradiction. EPR concluded that quantum mechanics should be an incomplete theory.

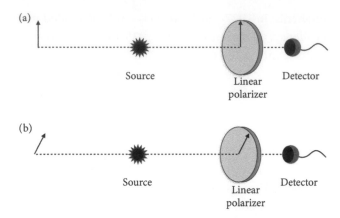

Fig. 12.1 Two photons initially prepared in an entangled state are sent in opposite directions. Alice chooses the polarization orientation in (a) along the y-axis and in (b) along an axis making 45° with the y-axis.

Another way to see the EPR argument clearly is through the simple two-state example of photon polarization.

Let us consider two photons A and B that are prepared in an entangled state: If photon A is horizontally polarized then photon B is also horizontally polarized and if photon A is vertically polarized then photon B is also vertically polarized. Let us suppose that they move in opposite directions and are very far away from each other. Let them be so far away that that they cannot influence each other in any way.

What we learnt in our discussion of entanglement is that, if Alice measures the A photon via a polarizing beam splitter in the orientation $\oplus$, and if she finds the photon to be vertically polarized, $\uparrow_A$, then she can be certain that if Bob measures the B photon with the same setting of the beam splitter, he will find his photon to be vertically polarized, $\uparrow_B$ (Fig. 12.1a). And if she finds the photon to be horizontally polarized, $\rightarrow_A$, then she can be certain that if Bob measures the B photon with the same setting of the beam splitter, he will find his photon to be also horizontally polarized, $\rightarrow_B$.

However, if Alice measures the A photon via a polarizing beam splitter in the orientation $\otimes$, and if she finds the photon to be polarized along

45 degrees, $\nearrow_A$, then she can be certain that if Bob measures the B photon with the same setting of the beam splitter, he will find his photon to be polarized along 45 degrees, $\nearrow_B$ (Fig. 12.1b). Similarly, if she finds the photon to be polarized along 135 degrees, $\nwarrow_A$, then she can be certain that if Bob measures the B photon with the same setting of the beam splitter, he will find his photon to be also polarized along 135 degrees, $\nwarrow_B$.

Now EPR's argument runs as follows:

Alice makes a measurement on photon A by passing the photon through a polarizing beam splitter with $\oplus$ orientation. If the photon is found in state $\uparrow_A$, she can conclude instantaneously that photon B's polarization is also along the vertical direction, i.e., $\uparrow_B$. Alice made this conclusion "without in any way disturbing" the photon B. Thus, according to EPR, there should be an element of reality associated with the polarization $\uparrow_B$ of the photon B.

If, on the other hand, Alice chose to make a measurement on photon A in the orientation $\otimes$, let her outcome be $\nearrow_A$. She then concludes that the photon B's polarization should be $\nearrow_B$. As before, Alice makes this conclusion "without in any way disturbing" the photon B and according to EPR, there should be an element of reality associated with the polarization $\nearrow_B$ of the photon B.

We thus have a situation where it is possible to determine the state of photon B in the two orientations $\oplus$ and $\otimes$ of the polarization beam splitter and these polarization states are obtained "without in any way disturbing" photon B. Then, EPR argue that, in any complete theory, "every element of the physical reality must have a counterpart." Therefore, the states of the polarization, $\uparrow_B$ and $\nearrow_B$ for photon B should be obtainable from quantum mechanics if it is a complete theory.

This is a paradoxical result as quantum mechanics is unable to determine the states of polarization of a photon in the two bases $\oplus$ and $\otimes$ simultaneously. This point has been discussed in detail in Section 6.2. Basically, if we measure the polarization in the basis $\oplus$ as $\uparrow$, then the outcome in the basis $\otimes$ is completely unpredictable. There is 50 percent probability each for the outcomes $\nearrow$ and $\nwarrow$. Same is true if the polarization of a photon in the polarization state $\nearrow$ is measured in $\oplus$. There is a 50 percent probability each for the outcomes $\uparrow$ and $\rightarrow$. We cannot assign the polarization of a single photon in both bases

$\oplus$ and $\otimes$ simultaneously. This result is in line with Bohr's principle of complementarity.

This result has come to be known as the "EPR paradox." The inability of quantum mechanics to make definite predictions for the outcome of certain measurements led EPR to conclude that quantum mechanics is an incomplete theory. They postulated that the theory may be made complete if we include certain "hidden variables" that are not known and perhaps not measurable. It was hoped that an inclusion of these "hidden variables" would restore the completeness and determinism to quantum mechanics.

It should also be noted that the locality condition is central to the EPR argument. It is important that Alice's choice of the basis should not influence the outcome of the measurement at Bob's end in the same basis. Here we elaborate this important point.

Let us first consider the situation when the locality condition is not satisfied. Both Alice and Bob are close enough that the choice of the basis for measuring Alice's photon is known to Bob. Then it is quite possible that the photons at Alice's and Bob's ends can communicate. If Alice chooses the $\oplus$ basis, then Bob's photon knows the orientation in which it will be measured. Thus, if Alice's outcome is vertical polarization, $\uparrow$, then Bob's photon which was correlated at the moment they were created, would also be found with vertical polarization $\uparrow$. Similarly, if Alice chooses the $\otimes$ basis and finds the photon polarized along 45 degrees $\nearrow$, Bob's photon will be found in the $\nearrow$ polarized state. In this case we can argue that the photons have only a single direction of polarization, either along 0 degree or 45 degrees. They choose the appropriate direction depending on the orientation of the beam splitter. Thus, when the locality condition is not satisfied, it should be possible to explain the experimental results, at least in principle.

However, when the locality condition is satisfied, Alice and Bob are very far from each other, and light may take a long time to reach from Alice to Bob. Now Bob's photon does not know the orientation of Alice's beam splitter. In this case, if Alice's photon is measured in the $\oplus$ basis and found with vertical polarization $\uparrow$, Bob's photon will also be found with vertical polarization $\uparrow$. Also, if Alice's photon is measured in the. $\otimes$ basis, then, if Alice's photon is found with polarization in the polarization state ↗, Bob's photon will also be found with polarization state ↗. The crucial difference in the two situations, when the locality condition is not satisfied and when it is satisfied, is that, in the second situation when the locality condition is satisfied, Bob's photon must have both polarizations (↑ and ↗) *simultaneously* present. Bob's photon should be ready to exhibit either polarization direction depending on Alice's choice far away. However as discussed above, quantum mechanics does not allow simultaneous reality of polarizations in the bases ⊕ and ⊗.

12.3 Bohr's reply

The EPR paper was published in 1935. This work represented a frontal attack on Bohr's principle of complementarity and the foundations of quantum mechanics. If the EPR argument is true then the principle of complementarity will no longer be true.

Leon Rosenfeld has recorded Niels Bohr's reaction to the EPR paper in these words:

> ...This onslaught came down upon us as a bolt from the blue ... Its effect on Bohr was remarkable. As soon as Bohr had heard my report of Einstein's argument, everything else was abandoned, we had to clear up the misunderstanding at once ... Bohr, greatly excited, immediately began to dictate a reply to Einstein. He found, however, that this was no easy matter. He'd start off on one track, then change his mind, backtrack, and start again. He couldn't put his finger where the problem was. "What can they mean? Do you understand it?", he would ask ... After some six weeks of work, Bohr had an answer ...

In his reply (also published in 1935 and under the same title as the EPR paper), Bohr argued that the reality of a property of an object does not exist until either it has been measured or is in a position to be measured with a predictable result. So, in Einstein's proposed experiment discussed above, the reality of the position of the second particle cannot be established until the position of the first particle is measured. In order to establish the reality of the speed of the second particle, a measurement of the speed of the first particle is needed. However, if the position of the first particle is measured precisely, the speed of the first particle becomes disturbed according to Heisenberg's uncertainty relation. The second measurement of the speed of the first particle is therefore made on the disturbed system. Therefore, the speed of the second particle cannot be considered precise. Bohr's principle of complementarity is therefore intact.

According to Bohr, quantum mechanics only provides us with a set of rules regarding the outcome of the measurements on the physical properties of the object when the measurements are carried out. In his words,

> The extent to which an unambiguous meaning can be attributed to such an expression as "physical reality" cannot of course be deduced from a priori philosophical conceptions, but ... must be founded on a direct appeal to experiments and measurements.

Bohr's case against Einstein rests on the premise that a measurement inevitably disturbs the system. The key point of Bohr's reply is that, in Einstein's argument, the two particles cannot be treated as independent until a measurement is made. Bohr argues that the inevitable disturbance during the measurement process leads to "final renunciation of the classical ideal of causality and a radical revision of our attitude toward the problem of physical reality." Reality at the quantum level does not exist until the object is measured.

We can illustrate this important point further by considering the two-state example of polarization again. We recall that a photon polarized at 45° in state ↗, when passing through a polarizing beam splitter with orientation ⊕, has equal probability of passing through in states → and ↑. Suppose it is found in state ↑. At this point we can assign the reality of the state ↑ to the photon. This means that if we make any further measurement on this photon in the ⊕ basis, we can say with certainty that the outcome will be ↑. But what about the measurement of this photon in the ⊗ basis? We have seen several times in our earlier discussions that the outcome will be ↗ and ↖ with equal probability. Suppose the outcome is ↗. At this time, we are again tempted to assign a reality of state ↗. Any further measurement on this photon in the ⊗ basis would yield the outcome ↗ with certainty. By now we are confident that the state ↑ and ↗ are true and real properties of the same photon. Next, we take the same photon which is in the state ↗ and pass it through the beam splitter with orientation ⊕. What should we expect? If the state ↑ represents the real property of the photon, then we should be confident that the outcome will be the state ↑. But this is not what happens. There is a 50 percent probability that it will be found in state →. This negates our common-sense conclusion that the state ↑ represented the real property of the photon.

This simple example helps to address the question of *reality* of quantum objects. What we see is that, as stated above, reality does not exist until the object is measured. Before the measurement, we cannot, in general, assign a definite or real state of the polarization to the photon. The role of the measurement device (a polarization beam splitter in the present case) is also very mysterious. How does it decide which state → or ↑ to select if the photon in the state ↗ passes through the beam splitter with orientation ⊕? This measurement problem is discussed in the next chapter. Here we note that Bohr's point of view, which has come to be accepted by the majority of physicists, is that there is no reality to the properties of quantum objects like the polarization states of a single photon, or the position and the speed of a particle. These properties become real only after they are measured. Some have gone as far as to say that physical objects like photons and electrons and the ball and the moon ... exist only when observed. They have no element of reality when they are not being observed. The moon indeed does not exist when no one looks at it. This is truly stunning.

In a nutshell, there is a fundamental difference in the definitions of reality by Einstein and Bohr. According to Einstein, a property of an object is real if it exists independent of a human observer. On the other hand, according to Bohr, we cannot assign reality to a property such as the position or the speed of an object until either it is measured or is in a position to be measured with a predictable result.

In the absence of a concrete experimental situation to test the *reality* and *locality* aspects of quantum mechanics, the debate concerning the foundations of quantum mechanics remained philosophical. The ultimate answer came about 30 years later, almost ten years after Einstein's death.

12.4 Bell's inequality

In 1964, John Bell proposed an inequality that would be satisfied by any theory whose main ingredients are *locality* and *reality*. This inequality involved experimentally measurable quantities and thus afforded an opportunity to experimentally test this inequality and see if our common-sense belief in locality and reality is justified. Quantum mechanics predicted that the Bell inequality would be violated.

The experiments were carried out in the 1970s and the results were in complete agreement with the predictions of quantum mechanics, violating Bell's inequality and thus proving that *reality* and *locality* do not coexist. This is an amazing result. How do we reconcile with it? Our common sense does not appear to be in conformity with the truth about reality and locality. Maybe there is no such thing as *reality*—the moon really does not exist when we do not look at it. Or the world is *non-local*—what we do here on earth now, it affects the stars and galaxies many trillions of kilometers away instantly. Or both.

The amazing result that has shaken our perception of nature and deeply influenced our concept of reality can be obtained in a highly mundane manner.

Let us assume that Alice and Bob live very far away from each other in a very different kind of climate. They live so far away that the weather at one place cannot influence the weather of the other place. We consider three characteristics of the weather in both places on a given day—it can be sunny or not sunny (we avoid the word cloudy here), it can be cold or not cold (again we avoid using the word warm), and windy or not windy. As example, Alice's place can be sunny, cold, but not windy one day and it can be not sunny, not cold, and windy on another day.

Let us also assume that the weather at the two places has a very strange feature: If it is sunny weather for Alice, it is not sunny (cloudy) for Bob and vice versa. Similarly, if it is cold for Bob, it is not cold (warm) for Alice and so on.

Table 12.1 The weather conditions observed by Alice. The weather conditions at Bob's end are just the opposite.

Day 1	sunny	cold	not windy
Day 2	sunny	not cold	windy
Day 3	not sunny	cold	not windy
Day 4	sunny	cold	windy
Day 5	not sunny	not cold	windy
Day 6	sunny	not cold	not windy
Day 7	not sunny	cold	windy
Day 8	sunny	not cold	not windy
Day 9	not sunny	cold	not windy
Day 10	sunny	cold	windy
Day 11	sunny	not cold	windy
Day 12	not sunny	cold	windy
Day 13	sunny	not cold	windy
Day 14	sunny	cold	not windy
Day 15	not sunny	cold	windy
Day 16	sunny	cold	windy

Let Alice and Bob observe the weather on a large number of days and compile the results. The possible weather conditions for Alice for sixteen days are given in Table 12.1.

Here it is understood that Bob's weather is just the opposite of Alice's weather. For example, on the first day, when Alice's weather was sunny, cold, and not windy, Bob's weather was cloudy (not sunny), not cold (warm), but windy. Thus, if we want to count the number of days when Alice finds her day sunny and Bob finds it cold, this corresponds to the days when Alice's day is sunny but not cold. For this counting we are not concerned about whether the day is windy or not. Thus, the number of days when it is sunny for Alice and cold for Bob are all those days in the above list when Alice's day is sunny, not cold, and windy as well as sunny, not cold, and not windy. These are days 2, 6, 8, 11, and 13.

With these considerations, the number of days that Alice finds sunny and Bob finds cold is five. Similarly, the number of days that Alice finds sunny and Bob finds windy are four (days 1, 6, 8, and 14) and the number of days Alice finds cold and Bob finds windy are also four (days 1, 3, 9, 14). It is clear that the sum of the days when it is sunny for Alice and cold for Bob (five) and when it is cold for Alice and windy for Bob (four) is greater than the number of days when it is sunny for Alice and windy for Bob (four).

Let us label the number of days when it is sunny for Alice and cold for Bob by N(sunny, cold). Similarly N(cold, windy) and N(sunny, windy) designate the number of days that are cold for Alice and windy for Bob and sunny for Alice and windy for Bob, respectively. In terms of these notations, the inequality becomes N(sunny, cold) + N(cold, windy) is greater or equal to N(sunny, windy). We can construct any table of our liking. This result will always be true. As we see below, this is a sort of Bell's inequality.

This inequality can also be derived pictorially as shown in Fig. 12.2. There are eight possible weather conditions: (s, c, w), (s, c, nw), (s, nc, w), (s, nc, nw), (ns, c, nw), (ns, c, w), (ns, nc, w), (ns, nc, nw). Here (s, c, w) denotes the day when Alice had sunny, cold, and windy weather whereas Bob had not sunny, not cold, and not windy weather. Similarly (s, c, nw) denotes the day when Alice has sunny, cold, and not windy weather whereas Bob has not sunny, not cold, and windy weather. The corresponding number of days are N(s, c, w) and N(s, c, nw) etc.

In Fig. 12.2, we represent the relative number of days with various weather conditions at Alice's end. Bob's weather conditions are just the opposite to those of Alice. The regions 1, 2, 3, 4, 5, 6, 7, and 8 are proportional to the number of days with weather conditions (s, c, w), (s, c, nw), (s, nc, w), (s, nc, nw), (ns, c, nw), (ns, c, w), (ns, nc, w), and (ns, nc, nw), respectively.

At this point we note that the number of days when the weather is sunny for Alice and cold for Bob, N(sunny, cold), is given by the number

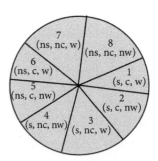

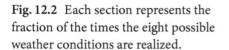
Fig. 12.2 Each section represents the fraction of the times the eight possible weather conditions are realized.

of days N(s, nc, w) plus N(s, nc, nw). This corresponds to regions 3 and 4 in Fig.12 2. Similarly the number of days where the weather is cold for Alice and windy for Bob, N(cold, windy), is given by regions 2 and 5 and the number of days that are sunny for Alice and windy for Bob, N(sunny, windy), is given by regions 2 and 4. This clearly shows that N(sunny, cold) + N(cold, windy) (equal to regions 3 + 4 + 2 + 5) is greater or equal to N(sunny, windy) (equal to regions 2 + 4). This is the same inequality that we mentioned above.

At this point, the reader may be wondering what all this discussion about the weather conditions at Alice and Bob's ends have to do with the questions of reality and locality. To answer this question, we note that only two concepts entered into our derivation of the inequality: N(sunny, cold) + N(cold, windy) is greater or equal to N(sunny, windy). These concepts are *reality* and *locality*. Reality enters into our analysis via an implicit assumption that the three weather conditions (sunny, cold, windy) and their opposites are real weather conditions; namely, they exist at Alice and Bob's locations irrespective of whether we observe them or not. Locality implies that what weather conditions Alice decides to look at are completely independent of the weather conditions that Bob decides to measure. A violation of this inequality would amount to saying that reality and locality cannot co-exist.

Next, we analyze this question within the framework of quantum mechanics. For this purpose, we consider an experiment with photons as shown in Fig. 12.3. Here a source generates two photons, one moving to the right to Alice and the other to the left to Bob. The photons are entangled in such a way that their polarizations are always orthogonal to each other. For example, if Alice's photon is horizontally polarized then Bob's photon is vertically polarized and vice versa. Similarly, if Alice's photon is polarized along 45 degrees, then Bob's photon is polarized along 135 degrees and so on.

In the experiment, Alice and Bob are separated by such a large distance that they cannot influence each other in any way whatsoever. This ensures that the *locality* condition is satisfied—the decisions made by both Alice and Bob do not affect each other. We also assume that each photon has a well-defined polarization no matter what the orientation of the polarizing beam splitter. For example, if the polarizing beam splitter

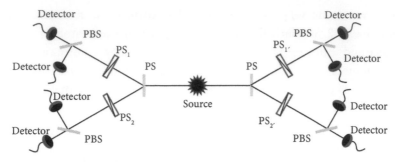

Fig. 12.3 Schematics for an experiment to test Bell's inequality. Here the source of light produces two photons. One goes to the left and the other to the right. The role of the path selectors PS is to direct the respective photon to upper or lower polarization beam splitters. An optical device called a Pockel cell PS_i chooses the basis in which the polarization of the photon is detected.

is set along the horizontal direction, then the photon will be polarized either in the vertical direction or the horizontal direction. Similarly, if the polarizing beam splitter is rotated by 45 degrees, the photon polarization will be along 45 degrees or 135 degrees, and so on. Thus, even if we cannot simultaneously measure the polarization along two different directions, we assume that the photon has a *real* polarization associated with the setting of the polarizing beam splitter. Thus, the *locality* and *reality* conditions are embedded in our analysis of the experiment.

We consider, both for Alice and Bob, the polarization measurement along three orientations of the beam splitter, along 0 degrees, 22.5 degrees, and 45 degrees. These orientations are chosen, for each photon pair, completely randomly and without any prior communication between Alice and Bob. The orientations are similar to the three weather conditions, namely sunny, cold, and windy, in the above example. The situation is completely analogous. Here the measurement of polarizations along 0 degrees, 22.5 degrees, and 45 degrees are similar to sunny, cold, and windy weather conditions. Like before, we count the number of the three kind of events: When Alice detects her photon along 0 degrees (horizontal) and Bob detects along 22.5 degrees; Alice detects her photon along 22.5 degrees and Bob detects along 45 degrees; and

Alice detects her photon along 0 degrees and Bob measures along 45 degrees. And as before, we expect that the sum of the number of events $N(0, 22.5)$ and $N(22.5, 45)$ should be equal to or larger than $N(0, 45)$. This is Bell's inequality that is based on only two premises: *locality* and *reality*. A violation of this inequality would imply that locality and reality cannot co-exist.

The experimental activity to test Bell's inequality has an interesting and rich history. The first experiment was done by S. J. Freedman and J. F. Clauser at the University of California at Berkeley in 1972. The outcome was in violation of Bell's inequality and agreed with the predictions of quantum mechanics. In 1974, R. A. Holt and F. M. Pipkin carried out a similar experiment at Harvard University. This experiment was in agreement with Bell's inequality but in disagreement with quantum mechanics. This work was never formally published and remained in preprint form. So, in 1974, there were two experiments at the two coasts of the United States, both disagreeing with each other. A decisive experiment was carried out by E. S. Fry and R. C. Thompson at Texas A&M University two years later in 1976. This experiment proved Freedman and Clauser correct and demonstrated the violation of Bell's inequality. Another landmark experiment that explicitly addressed the issue of locality was done by A. Aspect, J. Dalibard, and G. Roger in 1982. Aspect, Clauser, and Anton Zeilinger received the 2022 Nobel Prize for Physics.

Since those early days, a large number of experiments have been performed on various systems to test Bell's inequality and all of them showed a violation of this inequality in agreement with the predictions of quantum mechanics. We thus have the result that locality and reality do not co-exist. This challenges our commonsense view of reality. We are forced to ask the question: Is the universe real or is it an illusion?

13

Single universe or multiple universes

There was a curious pleasure in making oneself believe that time and space are unreal, that matter is an illusion and that the world consists of nothing but mind.

– Bertrand Russell

The universe is like a vast ocean. Wherever we look, it seems to be extended to infinite distances. The sky is full of stars, trillions of stars making a galaxy, and billions of galaxies forming clusters of galaxies. This is not all. When seen with high power telescopes, there is direct as well as indirect evidence for many other exotic stellar objects. As we discuss in Chapter 14, there are billions of neutron stars with a density so large that one tablespoonful of such a star would weigh 4 billion tons. Neutron stars form the last dying stage of a once majestic star. There are also flickering pulsars radiating like cosmic light towers, and then there are trillions of the most exotic stellar objects of all, black holes. Black holes are point-size objects, smaller than the smallest object we can think of, yet billions of times heavier than the sun. The known laws of nature that appear to be valid everywhere in the universe cease to exist in the neighborhood of black holes.

The vastness of space makes us feel that the universe that we live in must be the only universe. Any suggestion that ours may be one of an almost infinite number of universes sounds preposterous. Even more unbelievable are the propositions that these large number of universes cannot interact with one another, and billions of new universes are being created each second.

This all appears to be the stuff of science fiction. However, there are reasons to believe that ours may not be the only universe and there are an unimaginably large number of universes.

What motivates us to think that our common-sense view of a single universe may be wrong and there are other universes? In order to answer this question, we need to go back to the laws of quantum mechanics. These laws are so strange and mysterious that a conjecture of the existence of multiple universes does not appear to be a big price to pay to bring sanity to them.

We have seen that, in spite of the great success of quantum theory in explaining the observed phenomena, the conceptual foundations are still murky. The notion of wave–particle duality, the probabilistic nature of the experimental outcome, the questions about reality are all very mysterious. Another important question relates to the measurement problem.

There is a consensus among physicists about the outcome of any experiment. However, there continues to be discussion about the interpretation of these results. These issues, related to the measurement problem, have led physicists to come up with different interpretations. Many possibilities are discussed. One of the most amazing resolutions of the measurement problem is via the possibility of the existence of multiple universes. This many-worlds approach is a desperate attempt to solve the measurement problem.

13.1 Quantum measurement problem

The basic measurement problem can be highlighted with our simple example of a photon passing through a polarization beam splitter.

As we noted before, a vertically polarized photon ($\uparrow$) can be treated as being in a coherent superposition of states with polarization along 45 degrees ($\nearrow$) and 135 degrees ($\nwarrow$). This is akin to saying that the photon exists simultaneously in states $\nearrow$ and $\nwarrow$. If this photon passes through a polarization beam splitter which is rotated by an angle of 45 degrees with respect to the horizontal and is detected, it is found either in the state $\nearrow$ or state $\nwarrow$ with equal probability. The superposition disappears and the photon polarization state *collapses* to one of the two states, $\nearrow$ or $\nwarrow$.

As another example, we consider an electron moving toward a fluorescent screen. According to quantum mechanical laws, the electron is described in terms of a wavefunction. The wavefunction that contains all the information about the electron evolves according to the Schrödinger equation. Just before hitting the screen, the wavefunction of the electron is distributed over the entire screen. The physical interpretation of the wavefunction, as we discussed in Chapter 5, is that the square of the wavefunction is proportional to the probability of finding the electron at any point on the screen. Quantum mechanics provides us only with the probability of finding the electron at a particular location on the screen. However, when the electron hits the screen, it is found at a single point, not smeared over the entire screen. Again, it appears that the wavefunction of the electron, which was spread out over the entire screen just before the measurement, *collapsed* at a single point instantaneously as a result of measurement.

The question is: How do we interpret the interaction of the photon or electron in the above examples with the measurement devices? This interaction of the particles with the measurement apparatus forms the measurement problem that has haunted physicists since the very early days of quantum mechanics. The problem is that quantum mechanics, particularly the Schrödinger equation, does not address the measurement problem directly. Quantum mechanics only describes the evolution of the wavefunction or the coherent superposition with time, but does not answer about the interaction with the measurement apparatus.

This measurement problem was highlighted by Schrödinger via his cat paradox. As long as the cat is sealed inside the box, she is in a coherent superposition of states in which she is alive *and* she is dead. However, as soon as we open the box, the cat is found either alive *or* dead, but not both. This seems quite strange. This notion of a cat, both alive and dead, becomes even more mysterious if we replace the cat by a human. This we do in a later section.

The measurement problem is how the *collapse* of the wavefunction takes place. At what moment does the cat change from the state in which she was both alive *and* dead to a state in which she is either alive *or* dead?

The measurement problem highlights both the deterministic nature of the Schrödinger equation and the probabilistic nature of the measurement. For any initial state of the system, such as an atom in a coherent

superposition of energy states, the wavefunction evolves determinis-
tically to another state as time progresses. The Schrödinger equation
determines the continuous evolution to another quantum superposition
of states at a later time. The mysterious part is the measurement process
in which the system in a coherent superposition of states "jumps" to a
single randomly chosen state. This process is discontinuous.

How does this happen? There are three components of the measurement process: the system that is to be measured, the measurement
device, and the observer. Let us illustrate with a simple example. Suppose we have a photon in a coherent superposition of polarization states
$\nearrow$ and $\nwarrow$. A simple measurement device could be an atom whose electron is initially in the lowest energy state G. Let us suppose that, if the
polarization state is $\nearrow$, the photon is absorbed and the electron jumps
to the excited state E and, if the polarization state is $\nwarrow$, the photon is
not absorbed and the electron stays in the state G. At this point we are
tempted to say that, if the electron is found in state E, then the photon was
in the state $\nearrow$ and, if the electron is found in state G, then the photon was
in state $\nwarrow$. We then conclude that the measurement on the polarization
state is done.

However, a careful consideration shows that we are still as unknowl-
edgeable about the polarization state of the photon as we were before the
measurement was made. The problem still remains how to find whether
the electron is in state G or E. The problem has shifted—instead of finding the polarization state of the photon, we now need to find the state of
the electron. However, in order to determine whether the electron is in
state E or in state G, we need another measurement process to observe
the system. This process can continue forever. A quantum system cannot therefore be observed by repeated interactions with measurement
devices. The quantum nature of the measurement device makes the
collapse of the wavefunction to a single state impossible.

But where is the problem? If we look at the argument carefully, we
realize that reversibility is the main issue: the photon in a superposition of the polarization states $\nearrow$ and $\nwarrow$ leads to the atom, acting as a
measurement device, in an equivalent superposition of the states E and
G which contains the same unmeasured original superposition of the
polarization states that we started with. Another measurement device is

needed to find whether the electron is in state E or G. The Schrödinger equation, like Newton's equation, is reversible—if we know the quantum state now, we can project back in time and find the state at an earlier time.

How can we come out of this quagmire? Somehow an irreversible process has to be included in the measurement process. The need of irreversibility can be simply seen by noting that if the polarization is measured in (say) the state ↗, it should become impossible to recover the original coherent superposition of polarization states ↗ and ↖. However, all quantum systems evolve in a reversible manner according to the Schrödinger equation. The question is how to achieve irreversibility?

An irreversible process is typically a classical process and the measurement device follows the rules of Newtonian mechanics rather than quantum mechanics. This is an unacceptable situation as we have been claiming all along that every object, as tiny as an electron and as big as a planet or a star, follows the rules of quantum mechanics and classical or Newtonian mechanics is just an approximation valid where quantum effects are extremely small.

A way to see how irreversibility is obtained from systems that are reversible is to consider the simple example of a drop of blue ink dropped in a glass of clean water. What we observe is that, after few minutes, the blue ink is uniformly spread and the color of the water becomes light blue. This observation is trivial but it contains an element of mystery. The behavior of the ink–water system is irreversible—the atoms of the ink drop spread and make the color of the water light blue but it never happens that the light blue color water converts into a blue ink drop and a glass of clean water. How can that be? The motion of each atom in the ink drop is reversible—if we could make a movie of the atomic motion, we would not be able to distinguish if the movie is run forward in time or backward in time. But collectively, all the atoms in the ink drop evolve in an irreversible manner—we would immediately recognize that the movie in which the light blue water converts into the ink drop and clean water is being run backward in time. This simple observation that irreversibility of a system of large number of particles results from particles whose own motion is reversible lies at the foundation of an important branch of physics called thermodynamics.

An irreversible measuring device can, in principle, be constructed by small bags containing the blue ink. Let us assume that when a vertically polarized photon hits the bag of ink, nothing happens. However, when a horizontally polarized photon hits the bag, the bag is opened and the ink is released into the glass containing clean water. The clean water turns light blue in an irreversible process. The measurement is complete. By looking at the color of the water in the glass, we can conclude with definiteness which photon, horizontally or vertically polarized, is received.

This motivates us to see the measurement process as irreversible—once the measurement is made, it becomes impossible to return to the state that existed before the measurement. In addition, the outcome of the measurement is not deterministic. Instead, it is probabilistic.

Another more practical measuring device can be made via the process of spontaneous emission. We recall from Section 9.2 that quantum fluctuations in vacuum can cause the excited state electron in an atom to decay to a lower energy state and spontaneously emit a photon. Spontaneous emission is an irreversible process—the emitted photon can go in any of the infinite number of directions and it is not possible that this emitted photon can come back and lift the electron from the ground state to the original state by being absorbed.

With this insight, we can design a detector atom by introducing a third level D in our description of the atom that acts as the measurement device. We assume that the level D lies somewhere between the levels G and E. If the photon is in the polarized state $\nwarrow$, the electron stays in the level G. However, if the photon is in state $\nearrow$, it is absorbed and the electron jumps from level G to level E. The electron in level E can then jump to the middle level D and emit spontaneously a photon in the sea of vacuum. The net result is that if the photon is in state $\nwarrow$, the electron ends up being in state G and if the photon is in state $\nearrow$, the electron is in state D. The measurement process is complete as the final electronic state contains the information about the state of the incident photon. This process is irreversible and can be realized by noting that the ability of the spontaneously emitted photon to jump from level E to level D is lost and the reversibility cannot be attained without recovering that photon.

Typically, the measurement devices are very big, consisting of trillions and trillions of atoms. Such a device is deemed to follow the laws of classical mechanics to a very good approximation. The measurement of a quantum system irreversibly changes the state of the measurement device. For example, a combination of devices like photomultipliers connected to a needle on a meter serves to measure the characteristics of a photon. The movement of the needle indicates the experimental outcome.

These arguments, though satisfactory to most physicists, are deemed problematic for the purists. Here we treated the spontaneous emission in the surroundings as an irreversible process. By neglecting the surroundings and just concentrating on the measurement device (the atom), we achieve an irreversible, unique, and random measurement. However, the whole system, the photon, the atom, and the surroundings, form a reversible quantum system.

To some, the irreversibility requires an interaction with something outside the physical world. What can that be? One candidate is consciousness. A measurement process is complete only when the measurement outcome is registered in the consciousness of the observer. An observer is conscious of his own state which is irreversible. Therefore, the registration of a result in the consciousness brings a "collapse" of the wavefunction. The role of the conscious mind in the measurement process was highlighted by John von Neumann and Eugene Wigner, two giants of twentieth century physics. In the words of Wigner, "It is the entering of an impression into our consciousness which alters the wavefunction ... It is at this point that consciousness enters the theory unavoidably and unalterably." We discuss Wigner's argument in Section 13.3.

13.2 Copenhagen interpretation

What is the conventional wisdom about the measurement problem?

As we have seen in the last chapter, the widely accepted view is the same as advocated by Niels Bohr and Werner Heisenberg. This view is described as the Copenhagen interpretation of quantum mechanics as it took shape during the interaction between these two founders of quantum mechanics at Bohr's institute in Copenhagen in the mid to late 1920s. According to this interpretation, the role of quantum mechanics is only to make predictions about the experimental outcome. These predictions are, in general, probabilistic or statistical. For example, if, in the above example about photons, the probability of finding the photon in either state ↗ or ↖ is fifty percent, then there is no definite prediction about the outcome for a single photon. However, if we make measurements on a large number of identically prepared photons, then we will find fifty percent of the photons in state ↗ and fifty percent in state ↖. The important point is that quantum mechanics does not deal with the measurement process.

In the Copenhagen view of quantum measurement theory, the coherent superposition is broken by an act of measurement. For example, a photon in the superposition of states ↗ and ↖ is measured using a polarizing beam splitter, the superposition disappears, and the photon is found either in state ↗ or ↖. If this procedure is applied then correct answers are obtained for all possible situations involving the polarization of a photon and the orientations of the measurement devices. This becomes the end of the story.

The Copenhagen view does not deal with the questions relating to how the measurement apparatus determines the outcome. This view also denies the existence of the reality of a particular quantum state before the act of measurement. This point is discussed in the previous chapter.

In the absence of a more concrete theoretical basis, the Copenhagen interpretation has been adopted by the scientific community at large. However, a minority still struggles to understand the paradoxical issues associated with this interpretation.

13.3 Role of consciousness—Wigner's friend

Eugene Wigner raised the measurement problem to a new domain by including the conscious mind of the observer. Here we review Wigner's argument.

Instead of Schrödinger's set-up with a radioactive source, a hammer, a bottle of cyanide, and a cat, we consider a simpler system. Let us consider the same experimental set-up as discussed above: a photon passing through a polarization beam splitter inside a closed compartment. It is also assumed that a friend of Wigner is also present inside the compartment observing the experiment (Fig. 13.1a).

First let us consider the simple experiment in a closed compartment in which a photon is generated by a source. The photon, after passing through a polarizing beam splitter, is seen by Wigner's friend. Let us assume that this friend wears glasses with a polarizer that allows only light polarized along 45 degrees ($\nearrow$)and does not see it if it is polarized along 135 degrees ($\nwarrow$).

Suppose a photon polarized along 45 degrees ($\nearrow$) passes through the polarization beam splitter rotated by 45 degrees. The photon is, of course, seen by Wigner's friend, assuming that he has very sharp eyes that can detect a single photon. She records the outcome in her mind as A. In another experiment, the photon in state $\nwarrow$ is sent through the same beam splitter. This time she cannot see the photon and registers this outcome in her mind as B. So, we have two situations: If the state of the photon is $\nearrow$, the friend's state of mind is A and if the state of the photon is $\nwarrow$, then her state of mind is B. So far everything has happened

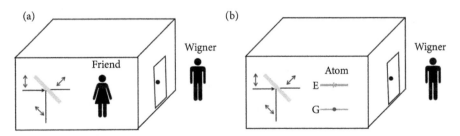

Fig. 13.1 A vertically polarized photon (↑) is incident on a polarizing beam splitter rotated by 45 degrees. (a) Wigner's friend sees the photon if it is polarized along 45 degrees ($\nearrow$), and does not see it if it is polarized along 135 degrees ($\nwarrow$). (b) Instead of Wigner's friend, an atom detects the photon. The atom remains in state G if the photon is in state $\nearrow$ and is in state E if the photon is in state $\nwarrow$.

in the closed compartment. At this moment, the compartment door is opened and Wigner asks his friend about her state of mind. If the friend's answer is A, Wigner concludes that the state of the photon's polarization was ↗. And if she answers B, the state of polarization was ↖. So far, there is no mystery!

Next, the same experiment is repeated with a photon polarized in the vertical direction (↑). This can be considered to be in a coherent super-position of states ↗ and ↖. The photon after passing through the beam splitter is found either in state ↗ or ↖. The corresponding state of mind is again A if the photon is detected by Wigner's friend in state ↗ and is B if the photon is not detected by her. The photon state, in this case would be ↖. Before opening the compartment, Wigner, sitting outside the compartment, concludes that the state of the photon and the state of his friend's mind are in an "entangled" state: if the state of the photon is ↗ then the state of mind is A *and* if the state of the photon is ↖ then the state of mind is B. When Wigner opens the door of the compart-ment and asks his friend about her state of mind and she says A, Wigner immediately concludes that the state of the photon was ↗. The state of the photon *collapsed* to ↗. Again, no mystery!

Now comes the paradox! Wigner asks his friend: what was the state of her mind before he opened the door of the compartment? His friend replies, it was A. So, the state of the photon had collapsed to ↗ before the door was opened and not after Wigner communicated with his friend.

The paradoxical situation arises if, instead of Wigner's friend, there was some other object that had no consciousness, like an atom. If the polarization of the photon was ↗, the electron inside the atom remained in one level (say G), and if the polarization of the photon was ↖, the elec-tron jumped to another level (say E). In this case, before Wigner opened the door, the joint photon–electron state was in an entangled state: if the electronic state was G, the photon state was ↗, and if the electronic state was E, the photon state was ↖. But, after Wigner opens the door and looks at the atom, if it is found in state G, the wavefunction collapse takes place and the photon is found in state ↗.

In the first case, when Wigner's friend was inside the compartment, the collapse of the wavefunction took place as soon as Wigner's friend detected the photon. This happened before the door was opened. However, in the second case, when an atom was inside the compartment, the collapse happened after the door was opened and Wigner looked at the state of the atom. In both cases, it was the measurement by a conscious person, Wigner's friend in the first case and Wigner himself in the second case, when the collapse of the wavefunction took place.

This is very mysterious. It raises an important question: Does measurement involve human consciousness? Wigner would claim that, in both cases, consciousness was responsible for the collapse of the wavefunction and the measurement. Consciousness seems to create the reality.

However, an immediate question is: What constitutes consciousness? Can a cat or a mouse, or even an insect create reality? And what happened when there were no life forms? Did the universe come into being only when the first life or the first human opened their eyes and looked up?

These questions concerning consciousness and its role in measurement are still discussed and debated. As, one of the foremost mathematicians and physicists of our era, Ed Witten, said in an interview,

> I think consciousness will remain a mystery. Yes, that's what I tend to believe. I tend to think that the workings of the conscious brain will be elucidated to a large extent. Biologists and perhaps physicists will understand much better how the brain works. But why something that we call consciousness goes with those workings, I think that will remain mysterious.

This is quite remarkable as it represents a scientist's admission about the limitation of science.

As discussed in the previous section, the Copenhagen interpretation has come to be accepted by most physicists. However, over the last century, many alternate interpretations of quantum mechanics have been presented and discussed. None has received as much attention as the many-worlds interpretation of quantum mechanics which seems to solve all these issues by postulating not one but many universes. However, it has its own problems. We present this esoteric picture in the next section.

13.4 Many-worlds interpretation

In 1955, Hugh Everitt, still a graduate student, presented a dramatic solution to the measurement problem that seemed to solve almost all the issues mentioned above. This solution gets rid of the nagging problem of the "collapse" of the wavefunction.

Let us first consider the Schrödinger's cat paradox. Until the box is opened, the cat is in a superposition state of being both alive and dead. According to the many-worlds interpretation, when the box is opened, if the cat is found alive, another universe is created in which everything is the same except that the cat is dead. We then have two universes which are identical in every respect except that, in one universe, the cat is alive and in the other the cat is dead. The two universes then continue to evolve until another measurement is made and new universes are created. Thus, in the process an almost infinite number of universes are created. These universes, however, do not interact with each other in any way.

This simple but really bizarre explanation of the measurement problem gets rid of the wavefunction collapse and is apparently internally consistent. In the many-worlds interpretation, every possibility is a reality. All the issues related to the questions of reality disappear in this simple and logical approach to the conceptual foundations of quantum mechanics.

This approach takes away the mystery from such notions as entanglement. If two photons are prepared in a superposition of states where the polarization of both photons are simultaneously oriented in the vertical direction and in the horizontal direction, upon measurement, we have two universes. In one universe, both polarizations are in the vertical direction and, in the other, they are in the horizontal direction. When the polarization beam splitter is rotated by 45 degrees, again two universes are created. In this case, both polarizations are along 45 degrees in one universe and along 135 degrees in the other universe.

The many-worlds interpretation on the one hand restores Newtonian determinism as we can predict the outcome precisely in different universes and, on the other hand, it is compatible with the probabilistic nature of quantum mechanics by giving a probabilistic outcome in the universe we live in.

There are, however, a number of issues with this interpretation. So far, we considered the examples with equal probabilities for the two choices. In this situation, it is easy to visualize the creation of two identical universes with two different possible results of the measurement outcome. But what about the situations when this is not true? For example, let us consider the case when, in the photon polarization experiment, the polarization beam splitter is rotated by 20 degrees with respect to the horizontal. In this case, for a vertically incident polarized photon, there is a 12 percent probability that the photon will emerge with a polarization along 20 degrees and 88 percent probability with polarization along 110 degrees. Here, it does not appear reasonable that the two universes are created with one in which the polarization is along 20 degrees and the other in which the polarization is along 110 degrees. How many universes are created in this case? If only two universes are created, then the result becomes identical to the case when the two outcomes in the output are equally probable. Scientists have tried to address this problem via more esoteric explanations.

There is another serious problem with the many-worlds interpretation. One of the cherished principles of physics is the conservation of energy. Energy can change character like mechanical energy can be converted into heat energy, but no new energy is created out of nowhere. As we learn later, according to Einstein's theory of relativity, energy can be converted into mass and vice versa. So, a more general principle is that the total mass and energy are conserved—energy can either be converted into another form of energy or converted to an equivalent mass. This principle does not appear to be holding in the many-worlds picture. An act of measurement produces an entire universe out of nowhere, a scenario in strict violation of the principle of conservation of energy.

Thus, in spite of solving most issues associated with the measurement problem, the many-worlds interpretation has many issues of its own. In view of our inability to test this hypothesis as the universes cannot interact with each other, the discussion of the many-worlds interpretation has entered into the realm of metaphysics. Most physicists continue to adhere to the Copenhagen interpretation as it works well unless we start asking philosophical questions.

PART 3
RELATIVITY AND COSMOLOGY

14

Stellar objects

Astronomy is older than physics. In fact, it got physics started by showing the beautiful simplicity of the motion of the stars and planets, the understanding of which was the beginning of physics. But the most remarkable discovery in all of astronomy is that the stars are made of atoms of the same kind as those on the earth.

– Richard P. Feynman

Since ancient times, humans have looked at bright objects in the sky and have wondered about their nature and movements. The earliest model was the geocentric model that established the centrality of the earth in the universe. Ptolemy standardized the model in the second century. According to the geocentric model, earth was believed to be at the center of the cosmos with seven planets orbiting around it. In order of increasing distance, they were the moon, Mercury, Venus, the sun, Mars, Jupiter, and Saturn. Later it was found that two objects on this list, the sun and the moon, are not planets. The seven objects were observable with the naked eye and existed within a background of a large number of stars. Ancient cultures saw patterns in the group of stars that resembled certain objects, animals, or mythical figures. The most famous of these constellations are Ursa Major (the great bear), Ursa Minor (the little bear), Orion (the hunter), Taurus (the Bull), and Gemini (the twins). Astronomy, the study of cosmic objects, is one of the oldest of sciences.

A major shift in our perception of the cosmos came when Nicholas Copernicus put forward the heliocentric model in 1543. This model placed the sun at the center with the planets, including earth, revolving around it. The centrality of the earth, and of humans, was gone.

The discovery of the seventh planet, Uranus, had to wait till 1781 when it became the first planet to be discovered using a telescope. The discovery of the next planet Neptune has an interesting story. This was the first planet that was predicted mathematically based on Newton's law of gravitation before it was directly observed in 1846. This was a stunning discovery that demonstrated the predictive power of a scientific theory. The last and the farthest planet, Pluto, was discovered by 23-year-old Clyde Tombaugh in 1930. Due to its rather small size, Pluto was downgraded to a "dwarf planet" in 2006.

No one knows who was the first to realize that the solar system and the stars are part of a galaxy called the Milky Way. The Milky Way is a streak of light spread through the sky when observed at night from a location far away from the city lights. Without a telescope, it is hard to distinguish individual stars within the galaxy. The first person to see that the Milky Way was made up of a very large number of stars was Galileo Galilei. He could see these individual stars when he pointed his telescope, one of the earliest, at the Milky Way.

Harlow Shapley, an American astronomer, made the first reliable measurements of the Milky Way in 1917 and concluded that the galaxy was planar with a hump in the middle. Since then, a large number of exotic stellar objects that are visible only through sophisticated telescopes have been discovered.

The earliest telescopes, going back to the early seventeenth century, helped astronomers to observe the stars using visible light. But as seen before (Chapter 3), objects can emit radiation over a very wide range of frequencies, from radio waves to gamma rays and X-rays. These frequencies are invisible to the naked eye. The first radio telescope was built in 1937, which enabled astronomers to see objects that were otherwise invisible. The first gamma-ray telescope, launched in 1961, helped in seeing exploding stars. Around the same time, telescopes operating in infrared opened up the possibility of exploring objects based on their heat emission. Our deep understanding of the cosmos, as will be discussed in Chapter 18, came through the measurement of the microwave radiation found throughout the universe. In 1992, NASA launched the Cosmic Microwave Background Explorer (COBE) satellite to probe the universe as it existed a very long time ago. In 1990, the first space-based optical telescope, the Hubble Telescope, was launched. This telescope, free of aberrations due to the earth's atmosphere, has provided the most detailed view of the universe. A much more powerful telescope, the James Webb Space Telescope, was launched by NASA in December 2021. The first image taken with the Webb telescope shows galaxy clusters as they appeared 4.6 billion years ago.

The most exciting results of the last century of astronomical observations are that we have been able to answer some of the most fundamental questions about the universe. How has the universe evolved? Why are stars, including the sun, so bright? What happens to stars when they get old? How big is the universe? The answers to these and similar questions, that are still being studied, have led to a revolutionary understanding of the cosmos.

14.1 Galaxies

The most familiar of all stars to us is the sun which is the source of light and heat. For us, the sun is the source of life, and therefore the most important stellar object. However, the sun is just one of the very large number of stars in our galaxy, the Milky Way. The sun and its planetary system are located at an insignificant position, about 30,000 light years out from the galactic center. Here, in order to avoid dealing with large numbers, the cosmic distances are not given in terms of the conventional units, miles or kilometers, but in terms of light years. A distance of one light year is the distance light at the speed of 300,000 kilometers per second travels in one year. In terms of kilometers this is given by 300,000 multiplied by the number of seconds in one year. This comes out to be about 9.5 trillion kilometers. The closest star, Proxima Centauri, is more than 4 light years, that is 38 trillion kilometers, away and the closest star with its own planetary system, Epsilon Eridani, is over 10 light years, that is 95 trillion kilometers, away. These are incredible distances.

According to the latest estimates, there are about 200 billion stars in the Milky Way whose radius is about 100,000 light years. The galaxy,

Fig. 14.1 An artist's rendering of the Milky Way and its central bar structure. The sun is located about 30,000 light years away from the center. There are about 200 million stars in the Milky Way that are spiraling around the center.

(Credit: NASA/JPL-Caltech/R. Hurt)

including the solar system, is spiraling around, completing one cycle in 200 million years. As such earth is orbiting around the sun and the sun is orbiting around the center of the Milky Way galaxy (Fig. 14.1).

There are trillions of galaxies in the observable universe whose radii range from 3000 to 300,000 light years. The nearest galaxy to our own, the Andromeda galaxy, is 2.5 million light years away. Galaxies range in size from dwarfs with just a few hundred million stars to giants with one hundred trillion stars, each orbiting its galaxy's center. There may be trillions of galaxies in that part of the universe that is inaccessible to us through telescopes and other means. More about them in the chapter on the big bang theory.

How did this galactic structure come about? This will also be discussed in the chapter on the big bang. Here we discuss the evolution of stars, with special reference to the sun (Fig. 14.2).

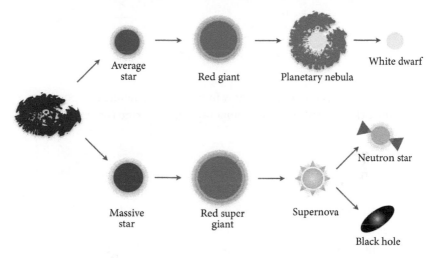

Fig. 14.2 The evolution of a star. The upper sequence corresponds to a small star and the lower corresponds to a big star, more than five to ten times the size of the sun.

14.2 Life of a star

A shining star mainly consists of hydrogen and helium atoms. The main source of energy in a star comes from the fusion process in which two hydrogen atoms fuse together and make a helium atom. The process takes place as follows.

We recall that the nucleus of a hydrogen atom consists of a proton which is surrounded by an electron. Sometimes two hydrogen atoms fuse together making a nucleus consisting of two protons surrounded by two electrons. Through a nuclear process, a proton converts into a neutron and a positron. A positron is the antiparticle of an electron which, when it combines with one of the electrons, both electron and positron are annihilated, and a high energy photon is released. The result is the formation of a special kind of hydrogen whose nucleus consists of one

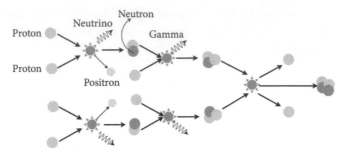

Fig. 14.3 The fusion process of how four protons generate two positrons and the helium nucleus. Here gamma rays and neutrinos are massless particles.

proton and one neutron. This is called deuterium. When two deuterium atoms fuse together, a helium atom is formed with a nucleus consisting of two protons and two neutrons. A somewhat more detailed description of the fusion process is shown in Fig. 14.3.

An enormous amount of energy is released in the process of nuclear fusion of hydrogen into helium. The reason is that the mass of the helium nucleus is less than the total mass of the two hydrogen nuclei. The surplus mass is converted into energy in the form of heat and light. The notion that mass can be converted into energy is discussed in the next chapter. This is the same energy that is released in the hydrogen bomb, and that we would like to harness as an energy source. In the sun, about 600 million tons of hydrogen fuses into helium every second. So, a good way to picture the interior of a star is to assume that a very large number of hydrogen bombs are exploding every second.

The fusion process in a star like the sun has been going on for about five billion years and is expected to continue for another five billion years. After that, the Sun will run out of the hydrogen in its core after it has been converted to helium, and the next phase will begin. At this point the star will start to collapse under its own weight. This process can be understood from Fig. 14.4.

In any star, there are two forces: the gravitational force tends to collapse the star into as small an object as possible while the heat produced in the core by the fusion process tends toward exploding the star. When these forces are about equal, the star remains stable. However, as the

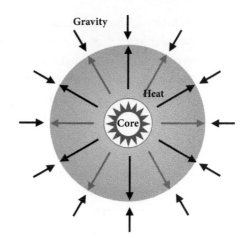

Fig. 14.4 There are two forces acting inside a star. The heat generated by the fusion process tends to explode the star but the gravitational force tends to collapse the star.

fusion process is depleted, the outer pressure is reduced and the star collapses under the gravitational force. The density of the core, and hence the temperature, undergoes a marked increase. However, the outer layers expand. In the case of the sun, the expansion of the outer layer will be so large as to engulf the nearest planets, Mercury and Venus, and to render the earth uninhabitable. The sun will become a red giant.

Red giant stars are not necessarily red. They can also have a range of colors from yellow and orange to various shades of red. In general, the term red giant star represents the late evolutionary stage of low-mass stars, whereas a red supergiant star is the late evolutionary stage of high-mass stars. Red giant stars are among the biggest stellar objects in the evolution of stars. After exhausting their supply of hydrogen, a process is initiated with their outer layers expanding considerably. The temperature of this outer layer can be as low as 5000 degrees. Canis Majoris is the largest red giant star at a distance of about 4000 light years away. It is 1.420 times the size of the sun with a mass equal to about twenty times the mass of the sun and a luminosity about 500,000 times that of the sun.

At this point, we make a distinction between a red giant and a red supergiant. Their subsequent evolutions are dramatically different as shown in Fig. 14.2.

For a low mass star like the sun, the red giant star stage may last for another billion years. During this phase, the star continues to shed a fraction of its mass into space. A disk of gas is formed around the core—a nebula is formed. The nebula consists of cosmic dust, hydrogen, helium, and molecular gases. Nebulae can be quite big, some of them spreading over several hundreds of light years. Their density is however very low—an earth-sized nebula cloud could weigh only a few kilograms.

The last stage of such a star is to become a white dwarf. When the nuclear fusion process is completely stopped, the core is left with helium fused into carbon and oxygen. The material in a white dwarf no longer undergoes fusion reactions. The star has no source of energy, and it starts cooling down eventually becoming a cold object not able to radiate. This is the end of the journey for a small star.

The evolutionary stages for a bigger star, bigger than five to ten times the size of the sun, are different, however. As soon as the nuclear fuel of the star is exhausted, the balancing outward force due to heat is removed and the star collapses immediately under the force of gravity. A shock wave propagates to the center and rebounds, blowing the star apart. This process happens within seconds. This is the stage when the star turns into a supernova.

Supernovae are among the most spectacular events occurring in the sky. Material equal to several solar masses is released at speeds up to several percent of the speed of light sweeping up a massive shell of gas and dust. Sometimes the expanding shock waves can trigger the formation of a new star. Supernovae can generate gravitational waves, though none has been detected so far.

Supernovae are a common occurrence in most galaxies, happening about once every fifty years (Fig. 14.5). So far more than 10,000 super-novae have been observed. However, disappointingly, they have all happened in other galaxies. The last time a supernova was observed in the Milky Way galaxy was in 1604. This supernova had exploded only 20,000 light years away from earth. It is still a mystery why no supernova has been observed in our galaxy since then. A supernova explosion close to earth in these times of highly sophisticated telescopes would be able to provide a wealth of valuable information regarding these powerful events.

The final stage is either to become a neutron star or a black hole depending upon the mass of the core. These are among the most exotic cosmic objects. Stars of the size of the sun become neutron stars, whereas

Fig. 14.5 A supernova appeared in the galaxy NGC 4526 in 1994.
(Credit: NASA)

stars of much bigger size collapse into a black hole. We devote the next section to describing the formation and properties of a neutron star and devote a whole chapter (Chapter 17) to black holes.

14.3 Neutron stars and pulsars

After a supernova explosion, the core keeps getting compressed under the action of gravity. The force becomes so large that the atoms lose their identity. We recall from Chapter 5 that an atom consists of a nucleus made up of positively charged protons and electrically neutral neutrons with negatively charged electrons surrounding the nucleus. Most of the mass (more than 99.9 percent) of the atom is concentrated in the nucleus whose size is only 1/100,000 of the size of the atom.

In the last stage of the life of a star, the compression force is so strong in the core that the electrons inside the atom are crushed into the nucleus. Inside the nucleus, these electrons and the protons combine together to make neutrons. What is left, in the place of atoms, is a massive object exclusively made of a very large number of neutrons. A neutron star is formed.

Except for black holes, neutron stars are the objects with the largest density in the universe—a neutron star of the radius of about 8 kilometers has the mass of the sun. A teaspoonful of the material from a neutron star would weigh two billion tons. A neutron star does not radiate light or heat. The life of a once majestic star comes to an end as a cold, dense remnant.

Neutron stars can exhibit many interesting features. The gravitation at the surface of the neutron star is about 200 billion times that of earth's gravity. In addition, the neutron star's magnetic field can be a trillion times stronger than the magnetic field of the earth. One of the most interesting discoveries in astrophysics was that of pulsars, which are particular examples of neutron stars.

When a rotating star collapses into a neutron star through a collapse of the core, the rotation rate increases substantially. This is due to the same effect as, when a skater rotating folds her outstretched arms, her rotation rate increases. Similarly, when a sphere is compressed into a sphere of reduced radius with the same mass, it rotates faster. The swiftly spinning cores of the collapsed stars with powerful magnetic fields produce jets of radiation that flash across the sky. If the magnetic poles do not coincide with the rotational axis of the neutron star, the beam of light sweeps the sky as shown in Fig. 14.6. If the earth is periodically in the path of the beam, pulses of light would appear, coming from a fixed point in space. This is like the rotating beam of a lighthouse. The objects, called pulsars, are among the most important astronomical finds of the twentieth century. Indeed, the discovery of pulsars by Jocelyn Burnell

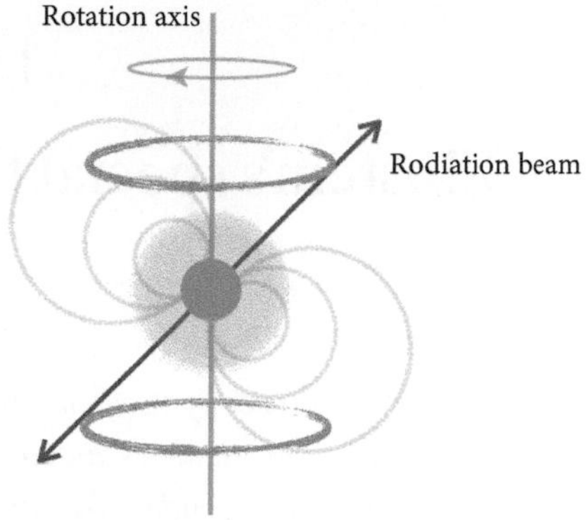

Fig. 14.6 An artist's view of a pulsar. The magnetic field is along an axis that makes an angle with the axis of rotation. The result is a beam of light that sweeps the sky.

and Antony Hewish in 1967 was the first observational suggestion that neutron stars exist. Hewish was awarded the Nobel Prize in Physics in 1974 for this discovery becoming the first ever Nobel laureate in the field of astrophysics.

15

Einstein's special theory of relativity

Henceforth space by itself, and time by itself, are doomed to fade away into mere shadows, and only a kind of union of the two will preserve an independent reality.

– Hermann Minkowski

In 1905, Albert Einstein published a paper that revolutionized our understanding of space and time. It dealt a fatal blow to the Newtonian laws of motion that were built on the assumption that space and time are independent of each other. The theory of relativity described objects moving at a constant speed with respect to each other. It explains how speed can affect time, space, and mass. For a moving object, the clock slows down, the length shrinks, and the mass increases. If the object is able to move close to the speed of light, time stops, the length shrinks to zero and the mass becomes infinite. The scenario is nothing short of science fiction.

But this is not all. Based on the premise that the speed of light is independent of the speed of the light source, the theory has led to the revolutionary idea that mass and energy are interconvertible. The development of the atomic and hydrogen bombs, and of course nuclear energy, are all consequences of this realization. This work helped in understanding the processes inside the sun that generate a huge amount of energy and is a source of life for us.

What motivated Einstein to present the theory of relativity? Let us see the conflicts between the laws of physics that Einstein wanted to resolve.

It has been well known since Galileo and Newton that it is impossible to decide whether an object like a carriage is moving in absolute space or at rest. For a person inside the carriage, is it the carriage or is it the person standing on the road or a platform that is moving? Indeed, no experiment is able to distinguish whether the carriage is at rest or is in a state of uniform motion (moving with constant speed in a straight line) with respect to the platform. This is according to the principle of relative motion. This principle can be stated as follows in a formal way: The laws of physics are identical to two persons in relative motion with constant speed. Einstein adopted it as the first postulate of his special theory of relativity.

In order to illustrate the relativity principle, let us consider a simple experiment. When a ball is dropped from a certain height, it bounces. If the surface of the floor is frictionless, then the ball comes back to the original height and continues to bounce. Next consider another person dropping the ball inside a carriage which is moving with a constant speed. What will he observe? He will observe the same thing—the ball bounces from the floor of the carriage. If the carriage is closed and no one can see outside, it would be impossible for someone inside the carriage to find out whether the carriage is at rest or moving with a constant speed.

Then there are the laws of motion as formulated by Newton that form the bedrock of classical mechanics. According to these laws, an object is accelerated if a force is applied, that is, as long as the force is applied, the speed of the object continues to increase. There is, however, no limit to how fast the object can be made to move. In principle, by applying a large force for a sufficiently long time, the speed can be increased beyond the speed of light in vacuum, 300,000 kilometers per second.

In 1862, James Clerk Maxwell published a paper in which he presented a set of equations that unified two forces of nature, the electrical force between charges and the magnetic force between the magnetic poles. These equations are called Maxwell's equations in his honor. A dramatic consequence of these equations was, as discussed in Chapter 4, that light could be described as an electromagnetic wave. The remarkable result was that Maxwell could derive an expression for the speed of light in vacuum in terms of some universal properties of electricity and magnetism. This implied that the speed of light must also be a constant of nature, just as the electrical and the magnetic field constants. In particular, the speed of light should not depend upon the speed of the light source such as a torch and remains the same no matter how fast the light source is moving.

This result was, however, in conflict with the Newtonian laws of motion. According to Newton's laws, the speed of light is dependent upon the speed of the light source. This is explained in the next section.

Therefore, a conflict existed between the laws of mechanics as formulated by Newton and the laws of electricity and magnetism as formulated by Maxwell. Only one of them could be right, surely? On the one hand, Newton's laws had stood the test of time for almost two centuries. On the other, Maxwell's equations were the culmination of a hundred-year effort to present the laws of electricity and magnetism in a unified manner. Giving up either one, Newton's laws or Maxwell's formulation, meant a giant step. This is the cross road that physics stood at when Einstein was brooding on these fundamental questions while serving as a Technical Assistant Third Class at the Swiss patent office in Bern in 1905.

Einstein, through his deep understanding and intuition, decided that the speed of light should be the same for all observers that are either at rest or moving with a uniform speed with respect to each other. He gave up Newton's mechanics in favor of Maxwell's conclusion about the speed of light. The importance of the realization that the speed of light does not depend on the speed of the light source cannot be overstated. This became the second postulate of Einstein's theory of relativity. This postulate shattered the Newtonian description of space and time as independent entities.

15.1 Speed of light

A common observation is that motion is relative. If we stand on the road and a carriage passes by at a speed of 40 kilometers per hour, then the relative speed between us and the carriage is 40 kilometers per hour. However, if we are in a car moving at 30 kilometers per hour in the same direction as the carriage then the relative speed between the car and the carriage will be 40 − 30 = 10 kilometers per hour. This means that if we crossed each other at a certain time, the distance between the carriage and car will be 10 kilometers after one hour, 20 kilometers after two hours, and so on. If the car also moves at 40 kilometers per hour, the same speed as that of the carriage, in the same direction as the carriage, then both the car and the carriage will be at rest with respect to each other. What this means is that, if at some time the two are separated by (say) 10 meters, they will remain separated by 10 meters after one hour or two hours or any other time. This is all according to Newton's laws of motion.

This behavior persists even when the speed is large. If a rocket is moving with a speed of (say) 10,000 kilometers per hour, we can get in a rocket moving at a speed of 15,000 kilometers per hour in the same direction as the first rocket. After some time, we will not only catch up with the first rocket, but get ahead of it. Again, no mystery! Everything is according to our expectations.

Next, consider a torch on the ground. When it is turned on, light rays are emitted which propagate at the speed of 300,000 kilometers per second. Thus, if we also stand on the ground, the speed of light is 300,000 kilometers per second. However, if we are in a rocket moving at a large speed of 10,000 kilometers per second in the same direction as the beam of light, the speed of light should appear to us to be $300,000 - 10,000 = 290,000$ kilometers per second and if our rocket is moving at 10,000 kilometers per second in the opposite direction, the speed of light should appear to be $300,000 + 10,000 = 310,000$ kilometers per second. Using the same logic as above, if we can build a rocket that moves at the speed of 300,000 kilometers per second in the same direction as the beam of light, the light, as seen from the rocket, will appear frozen.

Is this true? After all everything appears quite reasonable. Newton would also have told us that this should be true. But just to be fair to Newton, the speed of light was considered infinite before his life time. Ole Roemer, a Danish astronomer, who was Newton's contemporary, became the first person to prove that light travels at a finite speed. However, it took some time to accept the notion of a finite speed of light.

Einstein argued that if light followed the same rules as other objects, then it should be possible to move with the speed of light in a rocket in the same direction as the light signal and see the light signal at rest. According to Maxwell's theory, light consists of electric and magnetic waves (Section 4.7). These electromagnetic waves would then appear stationary with respect to him. He would see the electric and magnetic fields as hills and valleys just sitting like the wrinkles in a sheet. This would however contradict the work by Maxwell, whose equations did not allow such static fields in free space. According to Maxwell, light, consisting of electromagnetic waves, always moves at the same speed in a vacuum, 300,000 kilometers per second.

There were two theories that contradicted each other. According to Newtonian mechanics, we could move with the speed of light and the light would appear at rest to us. However, according to Maxwell, the speed of light is the same for all observers no matter how fast they are moving. Only one of them, Newton or Maxwell, could be right.

What happens if someone tells us that the rules of motion as discussed above are true for ordinary objects like trains and cars, but not true for light? This is exactly what Einstein did. Einstein agreed with Maxwell and concluded that light moves with the same speed whether the light source is at rest or moving with a constant speed.

This is the second postulate of Einstein's special theory of relativity, which, as seen below, had important, indeed revolutionary, consequences on our understanding of space and time. Einstein's theory of relativity was in direct contradiction to Newtonian mechanics, particularly when the speeds involved became comparable to the speed of light.

15.2 Michelson–Morley experiment

Toward the end of the nineteenth century, the concept of ether was firmly ingrained within the physics community. For example, Maxwell stated in an article entitled "Ether" for the Encyclopedia Britannica (1878):

> There can be no doubt that the interplanetary and interstellar spaces are not empty but are occupied by a material substance or body, which is certainly the largest, and probably the most uniform, body of which we have any knowledge.

He himself attempted unsuccessfully to measure the influence of ether's drag on the motion of the earth. It was, however, Albert Michelson and Edward Morley who carried out an experiment in 1887 with the objective of decisively establishing the existence of ether.

They sent light through a half-silvered mirror into an interferometer, now called a Michelson interferometer (Fig. 15.1). The light beam was split into two beams, one of them traveling straight to a mirror in one arm and the other propagating at right angles to another mirror. After reflecting from the mirrors, the two beams recombined at the beam splitter. They thus produced a pattern of constructive and destructive interference depending on the relative time light took to traverse the paths in the two arms. If the path difference was equal to zero or equal to a wavelength, then the two waves interfered constructively and a bright fringe was obtained. However, if the path difference was equal to one-half of the wavelength, then the two waves interfered destructively and canceled each other resulting in a dark fringe. Thus, a fringe pattern of alternately bright and dark fringes was obtained.

If the earth moves through an ether medium, the beam traveling along ether would take a longer time than the beam traveling in the perpendicular direction. Michelson and Morley expected a fringe shift equal to 0.4 fringes. What they measured was the maximum displacement of 0.02 and an average shift much less than 0.01. They thus concluded that the hypothesis concerning the existence of an ether medium is false.

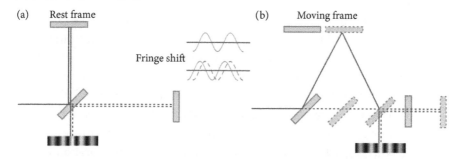

Fig. 15.1 Michelson interferometer set-up. A light beam from the left is incident on a beam splitter where part of the beam is reflected and part of the beam is transmitted. The reflected and the transmitted beams go to the mirrors and are reflected. These reflected beams are combined at the beam splitter and give rise to an interference pattern. (a) The interferometer is at rest and (b) the interferometer is moving through the ether causing the fringes to shift.

This null result—the most famous null result in the history of physics—was initially a major disappointment.

A resolution of the null result of the Michelson–Morley experiment came in 1889 by an Irish physicist, George FitzGerald. He postulated that the results of Michelson–Morley experiment could be explained using the hypothesis of the contraction of moving bodies in the direction of motion, the amount of contraction being just the right amount to give the same time difference as to explain the null result. According to him:

> I would suggest that almost the only hypothesis that can reconcile ... is that the length of the material bodies changes, according as they are moving through the ether or across it, by an amount depending on the square of the ratio of their velocities to that of light.

This was an amazing and highly counterintuitive resolution of the result of the Michaelson–Morley experiment. How can the length of an object shrink if it starts moving? But it turned out that he was on the right track. In 1892, unaware of FitzGerald's hypothesis, a Dutch physicist, Hendrik Lorentz, also came to the same conclusion. Today we call it a FitzGerald–Lorentz contraction.

Lorentz, however, went one step further. He derived them as a consequence of space-time transformation which was fundamentally different from a Galilean transformation. In the Galilean transformation, which formed the basis of Newtonian mechanics, time and space were completely independent of each other. Time flow was universal and did not depend upon where we were and how fast or slow we were moving. Lorentz showed that space and time were intertwined—if two persons move with respect to each other, for each one, the time duration and the length measurement for the other will be different. This was the first time that time and space were shown to be related and not independent of each other. This was, however, an ad hoc solution to the null result of the Michelson–Morley experiment with no justification based on a physics background. While Lorentz must be considered as the first to have found the mathematical content of the relativity principle, it was Einstein who stated the physical principles underlying relativity.

Einstein and Lorentz (1921)

Einstein formulated the theory of special relativity, based on the two postulates discussed above: The principle of relativity, i.e., the laws of physics do not change, even for objects moving in inertial (constant speed) frames of reference and the principle of the speed of light, i.e., the speed of light is the same for all observers, regardless of their motion relative to the light source. Based on these postulates, Einstein could derive the Lorentz transformation and the FitzGerald–Lorentz length contraction. This represented a major departure from Newton's notion of absolute space and time. The notion of stationary ether that had been the ever-existing background in all the theories since antiquity played no role in the theory of relativity.

15.3 Simultaneity

In Newtonian mechanics time was considered absolute and independent of space. Therefore, two events that are simultaneous for someone will be simultaneous for everyone, no matter where they are or what is their state of motion. For example, if two explosions took place in a big city at the same time, they will be regarded as simultaneous for everyone, not only in the same city, but for everyone in the world. This is true whether someone is sitting at home, walking on the street, driving in the car, or flying in an airplane.

An important consequence of Einstein's special theory of relativity is that space and time are no longer independent of each other, but are intertwined. A result is that the notion of simultaneity is no longer true for persons moving with respect to each other (for example, one person is walking on the street while another is driving past her). We illustrate this with an example which is due to Einstein himself.

Let us consider Alice standing on a railroad platform and Bob inside a carriage moving with certain speed on the train track. Let Bob be sitting right in the middle of the carriage. When the moving carriage is in front of Alice on the platform facing Bob, two firecrackers explode at the two ends of the carriage. We discuss these events both from Alice and Bob's point of view (Fig. 15.2).

For Alice, the two explosions take place at equal distance from her. The light from both the firecrackers will travel an equal amount of distance. After a time, during which the light travels the distance equal to half the length of the carriage, Alice will see the flash from both firecrackers simultaneously. She would conclude that the two explosions were simultaneous.

But what about Bob? Will the two events be simultaneous for him as well? Since the carriage is moving to the right with a certain speed, the light from the two firecrackers will travel different distances. When the light from the left firecracker travels the distance equal to half the length of the carriage, the carriage, and hence Bob, has moved farther. Therefore, light will have to travel a longer distance to reach Bob. However, light from the right firecracker will have to travel a shorter distance as Bob is moving toward the right firecracker. Thus, the light from the right firecracker will reach Bob earlier than the light from the left firecracker. Bob will conclude that the two explosions were not simultaneous. The right firecracker exploded before the left firecracker.

Thus, the two events (exploding firecrackers) are simultaneous to one observer (Alice) and non-simultaneous to the other (Bob).

This loss of simultaneity of the two events is a direct consequence of the special theory of relativity. In particular, it is based on Einstein's postulate that the speed of light is the same in all frames that are moving with constant speed with respect of each other.

It we analyze this experiment within the framework of Newtonian mechanics, the two explosions will be simultaneous for both Alice and Bob. In order to see this, we note that the speed of light from the two

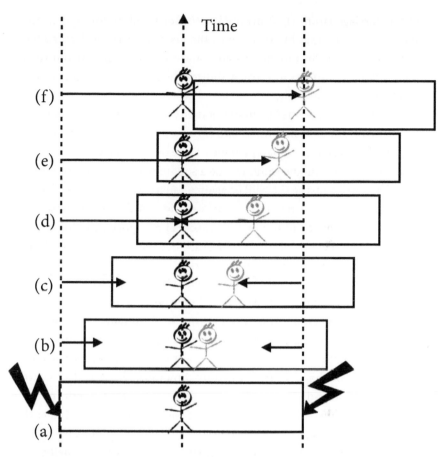

Fig. 15.2 (a) Alice and Bob are both at the same location. Alice is at rest and Bob is inside the compartment which is moving to the right. Two explosions take place at equal distance from both Alice and Bob. (b) Alice remains at rest and Bob moves to the right inside the compartment. (c) Bob sees the flash of light from the explosion on the right. Alice does not see light from either explosion. (d) Alice sees the flash of light from both explosions at the same time. (e) Bob continues to move to the right and the light from the left is moving towards him. (f) Light from the left explosion reaches Bob.

firecrackers would be different for Bob—the light moving in the direction of the motion of the carriage (from left cracker to Bob) will be higher and the light moving in the opposite direction to the motion of the carriage (from right firecracker to Bob) will be lower than the speed of light as seen by Alice. Since the faster light (from left cracker to Bob) has to travel a longer distance and the slower light (from right firecracker to Bob) has to travel a shorter distance, Bob would see the two lights simultaneously.

We can understand the loss of simultaneity from another point of view. In the three-dimensional world, two observers can see the same object differently if seen from different angles. For example, the view of a cubic box, when seen from the front, will be a square with a compressed third side. The view from the side will be again a square with compressed third side. The difference is that the compressed side for the second observer is the full-length side to the first observer. The important point is that an object in three dimensions can be described with differing lengths in different directions by different observers.

According to the special theory of relativity, space and time are no longer absolute—they describe a four-dimensional world, three dimensions in space and one time dimension. So, instead of an object in three-dimensional space, an event is described that happens at a given location in three-dimensional space at a given time. Let us consider two such events happening in the four-dimensional space. These two events in the four-dimensional space will have different lengths of any dimension for the two observers that are moving with respect to each other. The result is that they see the same event at different times. The interdependence of space and time explains the loss of simultaneity.

15.4 Time dilation and length contraction

The principle that the speed of light is the same for all observers moving with constant speed with respect to each other has another amazing consequence. To a person at rest, time passes more slowly for someone moving with a constant speed. For example, Alice standing on a platform would think that the clocks inside a carriage moving past her run slower. Here we illustrate this amazing result using a simple argument.

Suppose there are two identical clocks, one on the platform and the other inside the carriage. They work like this. A mirror is placed just above the standing person. A light signal is sent to the mirror, where it is reflected back. The time it takes for the light signal to travel to the mirror and back is defined as a redefined second.

For Alice standing on the platform, the redefined second will be equal to the duration for the light signal to travel to the mirror and back. This is equal to the distance travelled by the light signal, which is equal to twice the height of the mirror, divided by the speed of light. This is straightforward.

Next, we consider the duration of one second in the moving carriage. For Bob, inside the moving carriage, one second will again be equal to the travel time of the light signal to the mirror and back. The duration of one second will be the same for both Alice and Bob in their own locations.

But what about the duration of one second of Bob's clock as measured by Alice? From Alice's point of view, when the light hits the mirror, the carriage has moved a certain distance depending on the speed of the carriage. As seen in Fig. 15.3, the light signal has to travel a longer distance to reach the mirror. This is due to the motion of the carriage. When the light signal reaches the mirror, the carriage has moved, resulting in a longer distance than going straight up. Similarly, the reflected light has also a longer distance to travel to get back to Bob. The total distance traveled

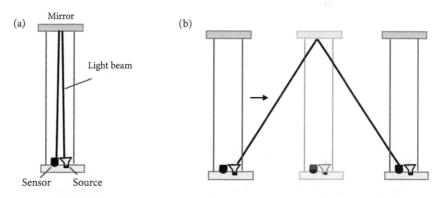

Fig. 15.3 A light signal is sent to the mirror and after a certain time the reflected signal is detected by the sensor. (a) For an observer at rest, the distance travelled by the light signal is smaller than (b) when an observer is moving with a certain speed.

by the light beam to go to the mirror and back will be longer. Since the speed of light is constant, the time duration corresponding to one second (light going to the mirror and back) will be larger. Thus, for Alice, the clock in Bob's carriage moves slower as compared to her own clock. For an ordinary speed of the carriage, the time difference may not be too big. However, if the carriage is moving close to the speed of light, the light beam will have to travel a much longer distance during one second as perceived by Alice. For Alice, the clocks in Bob's carriage must be running slower. How slow? Suppose Alice and Bob blink in one second. If Bob's carriage is moving at half the speed of light, an incredibly high speed, Alice would think that Bob took 1.15 seconds to blink. If Bob's carriage moved at 0.999 times the speed of light, Bob's blink would take 22 seconds and if Bob's carriage moved at almost the speed of light, say 0.999999 times the speed of light, Alice could take a 10-minute walk and Bob would still be blinking.

This effect is called "time dilation."

An amazing consequence of time dilation is that there will be disagreement between Alice and Bob on how much distance Bob travelled during a certain time, say one second. This point can be illustrated with an example.

Let us assume that Alice is standing on the platform whereas Bob is traveling in a carriage at half the speed of light, 150,000 kilometers per second. After a passage of one second, Bob will conclude that he has traveled 150,000 kilometers.

Alice, standing on the platform, knows that Bob's clock is slower. Bob's one second is equal to only 0.866 seconds as measured by Alice. Therefore, Alice would conclude that Bob has traveled a distance equal to the speed of the carriage, 150,000 kilometers per second, multiplied by 0.866 seconds. This would be equal to 129,900 kilometers. Thus, we have a highly counterintuitive result that the distance traveled by Bob in one second as measured by Alice is smaller than that measured by Bob.

Next, we ask a related question to both Alice and Bob: What is the length of the carriage that Bob is traveling in? Amazingly, Alice and Bob would give different answers. Again, let us illustrate with a simple example when Alice, standing on the platform, sees Bob, as before, traveling in a carriage at a speed of 150,000 kilometers per second.

First, from Bob's point of view, the carriage is at rest. It is Alice that is moving. How would Bob measure the length of the carriage? One way is to put a mirror at the farther end of the carriage. Bob sends a light signal from one end of the carriage which, after traveling the length of the carriage, is reflected by the mirror and travels back to Bob and is detected. If the total time taken by the light signal is 0.000002 second, Bob would conclude that the length of the carriage is the speed of light, 3000,000 kilometer per second multiplied by the total time of travel from him to the other end of the carriage and back, 0.000002 second, divided by 2. The division by 2 comes about because the total distance travelled is twice the length of the carriage. This is equal to 3 kilometers, indeed a very long carriage.

Next, we ask the question: What is the length of the carriage as measured by Alice who is standing on the platform? This is somewhat more complicated to answer, but let us try.

From Alice's perspective, the carriage is moving at half the speed of light. She would argue that Bob should have included the movement of the carriage in his calculation. According to her, light should take a longer time to travel to the mirror as the carriage (and the mirror) moved forward during the time the light signal travelled from Bob to the mirror. She also notes that light signal should take a shorter time to travel from the mirror back to Bob as Bob has moved forward by the time light signal reaches back to him. This is again due to the motion of the carriage. During the travel from Bob to the mirror, the time taken by the light pulse is equal to what it would take if the carriage was at rest *plus* an additional time to travel the distance that the carriage has moved. However, upon reflection back from the mirror to Bob, the time taken by the light pulse is equal to the time it would take if the carriage was at rest *minus* the time taken to travel the distance that the carriage has moved forward. Based on these observations, a simple calculation can be done. Alice would conclude that the length of the moving carriage is equal to 3/4 times the length of a stationary carriage. However, a further correction is needed. Alice knows that Bob's clocks are slower than her own clock. When this is included, Alice would conclude that the length of the carriage is $\sqrt{3/4}$ or 0.87 times the length of the carriage when it was at rest. This comes to 0.866 times 3 kilometers or 2.56 kilometers.

Therefore, Alice would see a smaller length of the carriage as compared to Bob. Here we considered the specific example of the carriage. However, this result is general and is true for all objects that are moving in the direction of the carriage.

The conclusion is that the length of the moving carriage appears to be shorter than the length measured by an observer at rest. If the carriage moves close to the speed of light, its length would approach to zero. This is called "length contraction."

An interesting observation is that the mathematical expression for the length contraction derived based on the postulates of the special theory of relativity agrees with the ad hoc expressions of the Fitzgerald–Lorentz contraction to explain the null result of the Michelson–Morley experiment. Here length contraction is derived as a consequence of the postulates of the theory of relativity instead of the movement of the interferometer in ether.

Time dilation and length contraction are truly bizarre consequences of Einstein's special theory of relativity. For an observer, Alice, looking at events and objects that are moving with a constant speed with respect to her, the progress of time has slowed down and the distances and size of the objects become shorter as compared to when they are at rest. The remarkable thing is that there is a symmetry—the observer in the carriage, Bob, would conclude that the progress of time has slowed down and the size of the objects has become shorter for events and objects on the platform. This conclusion is straightforward as the observer in the carriage considers himself at rest and sees the platform moving with the same speed in the opposite direction.

What is the evidence for these exotic effects?

An interesting and measurable evidence of time dilation and length contraction is found in the decay of an elementary particle called a muon. Earth's atmosphere is constantly bombarded with cosmic rays, mostly consisting of protons. These fast-moving particles collide with atoms in the atmosphere which disintegrate into some other elementary particles. One of these particles is the so-called muon. The muons, in turn, can decay into other particles. The decay process is usually probabilistic. The quantity that determines how long any particle stays without decaying is called its half-life. Half-life is defined as the time when half of the particles have decayed and the other half have not decayed. The half-life of muons is 1.56 microseconds. This means that if we start with one million muons then after 1.56 microseconds half a million muons will have decayed to some other particles and only about half a million muons will be left.

Muons are produced in the upper atmosphere at an altitude of about 10 kilometers and move at a speed of about 98 percent of the speed of light toward the surface of the earth. Experimentalists observe that about 5 percent of the muons (50,000 out of one million) produced in the upper atmosphere reach the ground and get detected.

Let us first see how these observations can be compared with the results obtained using Newtonian mechanics. The time taken for muons to travel 10 kilometers moving at 98 percent the speed of light will take 10 kilometers divided by 0.98 times the speed of light or 34 microseconds. The half-life of muons is 1.56 microseconds. Therefore, the time taken is 22 times the half-life of muons. This is an incredibly long time for any muon to be left undecayed. Less than one muon out of a million will reach the earth, in clear violation of the experimental result.

How does the special theory of relativity explain this result? This experiment becomes a testing ground of the consequences of the theory of relativity. We analyze the experimental outcome from two different angles.

First, we analyze it from the point of view of the scientist in a lab on earth. According to the time dilation for the relativistic muons, the time is dilated by a factor of 5. Thus, the half-life of muons that are moving close to the speed of light is not 1.56 microseconds but 5 times 1.56 microseconds or 7.79 microseconds. Thus, the muon undergoes 34/7.79 = 4.36 half-lives. It turns out that about 48,000 muons out of a total of 1 million muons should reach the earth. This is in agreement with the experimental results.

Next, we analyze from the point of view of the muon itself. The muon sees the earth moving toward itself with a speed of 0.98 times the speed of light. It sees the distance of 10 kilometers contracted by a factor of 5. The contracted distance is 10 kilometers divided by 5 or 2 kilometers.

The time taken by muons to travel this distance is 6.8 microseconds. The number of half-lives is equal to 6.8/1.56 = 4.36, the same result as before.

15.5 Paradoxes

A lack of simultaneity, time dilation, and length contraction are all highly counterintuitive and almost unbelievable. They shattered many ingrained ideas about space and time. Over the years many paradoxes have also been proposed as a result of this amazing but incredibly successful theory of physics. We present only a couple of them.

The first is the so-called "ladder in a barn" paradox. Here we illustrate it with an example (Fig. 15.4). Suppose Alice is standing next to a barn which is 4 meters deep. She sees Bob running toward her with a speed

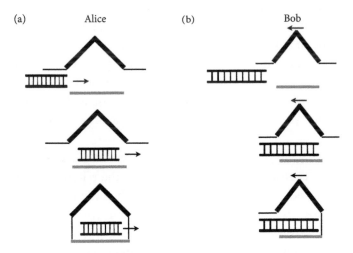

Fig. 15.4 The motion of the ladder and the barn from (a) Alice's point of view and (b) Bob's point of view. In (a), the barn is at rest and the ladder is moving from left to right at a speed of 0.8 times at the speed of light. Alice sees the length of the ladder reduced to 0.6 times its length at rest. In (b), the ladder is at rest whereas the barn is moving from right to left at a speed of 0.8 times the speed of light. Bob sees the length of the barn contracted to 0.6 times its length at rest.

of 0.8 times the speed of light. He carries a long ladder which is 5 meters long while holding it in the horizontal direction. The objective is to see whether the special theory of relativity allows the impossible: fitting the 5-meter-long ladder into a 4-meter-deep barn.

Alice standing next to the barn, sees Bob's ladder to be shortened by a factor of 0.6 due to length contraction. She concludes that the ladder is only 0.6 times 5 meters or 3 meters long. This is smaller than the depth of the barn which is 4 meters.

What is Bob's point of view? According to him, he is at rest holding a ladder that is 5 meters long. He sees the barn (and Alice) coming in his direction at a speed of 0.8 times the speed of light. To him the length of the barn is shortened and when he calculates the depth of the barn, it comes out to be 0.6 times 4 which is equal to 2.4 meters.

We thus have a paradoxical situation: Alice thinks the barn is 4 meters deep and Bob's ladder is 3 meters long whereas Bob thinks that his ladder is 5 meters long and the barn is only 2.4 meters deep. For Alice, the ladder fits comfortably inside the barn but, for Bob, the barn is simply too small to store the ladder.

Now Alice decides to do the following. When Bob, while running, enters the barn and hits the back wall, she closes the front door. Is the ladder inside the barn or not? According to Alice, the ladder should be inside the barn but Bob thinks this is impossible.

How to resolve this paradoxical result? The solution lies in the realization that the two events, Bob's ladder reaching the back wall and Alice closing the door, are not simultaneous for both Alice and Bob.

In Alice's frame, the two events, the ladder hitting the back wall and Alice's closing the door, are simultaneous. However, this is not true for Bob as he sees things differently. When the front end of the ladder hits the back wall, the back of the ladder is still outside the barn. The ladder hitting the back wall and the closing of the front door of the barn are not simultaneous for Bob.

Let us assume that the clock starts running when the ladder is at the entrance point of the barn. So after how much time will Alice close the door? The barn is 4 meters long and the ladder is moving at a speed equal to 0.8 times the speed of light. Alice will close the door at a time equal to the distance traveled divided by how fast Bob and the ladder are moving. This comes out to be $4/(0.8 \times c) = 5/c$ meters where $c = 300,000$ kilometers/second is the speed of light.

Bob's clock is slower than that of Alice. Therefore, a smaller amount of time must have elapsed for Bob. This is a bit complicated to calculate. The result from the theory of relativity is $3/c$ meters.

Next, we find how far the ladder has gone inside the barn. Since the relative speed of Bob and the barn is $0.8 \times c$, Bob moved inside by a distance of $(3/c)(0.8 \times c) = 2.4$ meters. This means that Alice wants to close the door when the front of the ladder is only 2.4 m inside the barn. This is much less than the 5 meters that Alice thinks the front of the ladder has moved. Alice and Bob do not agree when the door of the barn should be closed. This interesting argument solves the "ladder in a barn" paradox.

Perhaps the most celebrated paradox is the so-called "twin paradox." Again, we explain it via an example. We consider identical twins, named Alice and Bob. Alice stays on earth and Bob makes a round trip to the nearest star, Proxima Centauri, in a spaceship which is moving at a very high speed (0.995 times the speed of light). Let us examine the trip from the point of view of Alice and Bob.

From the point of view of Alice on earth, Bob travels to the star which is at a distance of 5 light years. It takes 5 years for Bob to reach the star and, moving at almost the speed of light, takes another 5 years to return to earth. She has aged 10 years when Bob returns to earth. What about Bob's age from Alice's perspective? Since Bob is traveling at 0.995 times the speed of light, the time is dilated by a factor of 10, and Alice concludes that Bob has aged only one year. Thus, if both Alice and Bob were 20 years old when Bob left on his trip, Alice has become 30 years old but Bob is only 21 years old at the time of his return.

What about Bob? What are the ages of Alice and Bob from his point of view? Bob remained at rest with respect to his spaceship throughout his journey. It was the earth, and Alice, that moved at 0.995 times the speed of light. He would conclude that Alice should be younger than him when he returns—a conclusion just opposite to that of Alice.

Who is older? Alice or Bob?

Before presenting the resolution of this paradoxical result, we present Einstein's own words in an article published in 1911: "If we placed a living organism in a box ... one could arrange that the organism, after any arbitrary lengthy flight, could be returned to its original spot in a scarcely altered condition, while corresponding organisms which had remained in their original positions had already long since given way to new generations. For the moving organism, the lengthy time of the journey was a mere instant, provided the motion took place with approximately the speed of light."

This result is puzzling only because we have presented it in such a way that the situation appears symmetric with respect to both Bob and Alice. For Alice, Bob is moving while, for Bob, Alice is moving. But this is not true. There are two ways in which the situation is asymmetrical. Alice has been at rest on earth during the whole time. However, Bob was mostly moving with a constant speed with respect to earth, but in order to return to earth, Bob has to reduce the speed of his spaceship at the end of his journey to Proxima Centauri, eventually stopping, turning around, and then accelerating to the same speed as in the outgoing journey to return to earth. The decelerating and accelerating parts of Bob's journey can cause further time difference. As we learn in the next chapter, for an accelerating object, time slows down. The situation for Alice and Bob is different from another angle. Bob's traveling trajectory has two parts: one going away from Alice and the other coming back to Alice. The paradox results as a consequence of this asymmetry. A careful analysis reveals that Bob will be older when he returns to earth.

15.6 Energy and mass are interconvertible, $E = mc^2$

In the previous sections, we have seen some incredible consequences of Einstein's theory of relativity. However, the most celebrated is the revolutionary result that energy and mass can be converted into each other via one of the most famous equations of physics, $E = mc^2$. Here E is the energy which is equivalent to the mass m times the square of the speed of light, c, where c^2 means c times c. Since the speed of light is huge, equal to

300,000 kilometer per second, c^2 is equal to a phenomenally large value – 9,000,000,000 kilometer per second × kilometer per second. Therefore, the amount of energy bound up even in a small amount of mass can be immensely large.

In order to appreciate the significance of this result, we recall that classical physics is built on certain conservation laws. For example, there are two separate conservation laws for energy and mass. According to these laws, in any process involving the interaction of objects either at rest or moving with respect to each other in the presence of any kind of forces, the total energy and total mass are separately conserved. Energy and mass cannot be created or destroyed. In any process, one form of energy can be converted into another form of energy but the total energy remains the same. The same is true of mass. What Einstein showed is that energy and mass can be converted into each other. The new law of conservation meant that the sum of the energy and the equivalent energy due to mass is conserved, but the individual conservation laws for energy and mass do not hold.

How to understand the equivalence of energy and mass?

A simple experiment can explain that. Consider a closed box as shown in Fig. 15.5a. Let there be a loaded pistol on the left side of the box which is attached to a timer. At a certain time, the pistol fires a bullet that travels to the right. At the moment the bullet is fired, the box recoils to the left.

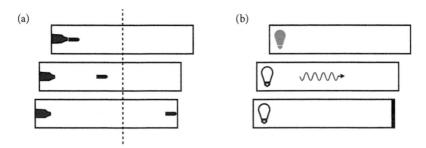

Fig. 15.5 (a) A gun is fired on the left side of the box which hits it and the bullet is embedded into the right wall. (b) A bulb sends a flash of light and the photons are absorbed by a dark paper on the right wall. The dotted line represents the center of mass.

This is in accordance with Newton's third law of motion that says that for every action there is an equal and opposite reaction. The bullet hits the right wall and is embedded in it. During the entire process the box is closed. The net result is that some mass (the mass of the bullet) has moved from the left to the right. As there is no external force on the box, the center of mass remains the same. Due to a redistribution of the mass (bullet moves from left to right), the box moves to the left. This whole experiment can be fully explained using Newtonian laws of mechanics.

Einstein's deep insight was to consider the same experiment as just described but, instead of a loaded pistol, there is a light bulb on the left side of the box. At a certain time, the bulb sends a flash of light to the right. The box recoils, though with a much smaller force. The situation is similar, when taking a picture, the camera is pushed backward when the flash gun is fired. The light pulse travels to the right and is absorbed by a dark piece of paper.

Let us assume, for the sake of argument, that the mass of the bullet is so small that the movement of the box to the left is the same in the two cases. Now Einstein reasoned as follows. If we do not see inside the box, we cannot distinguish the two situations. We would not know whether the box moved because the mass (of the bullet) transferred from left to right or the light pulse (which does not have mass but only energy) travelled from left to right. This clearly shows the equivalence of mass and energy. A careful analysis of this simple experiment leads to the most famous equation of the twentieth century, $E = mc^2$.

An important consequence of the mass–energy equivalence is that the mass of the object appears to increase with speed. The faster it moves, the "heavier" it appears to be. In our everyday life we do not observe this effect as most of the objects around us move with speeds that are much smaller than the speed of light. The increase in mass becomes perceptible when it moves at a significant percentage (about fifty percent) of the speed of light. Near the speed of light, the mass of the object appears to be infinite. Just like time dilation, the increase of mass is not felt by the object itself when it is moving. The increase of mass of the moving object is felt by observers who are themselves at rest.

A way to understand this counterintuitive result is to first realize that, when the object is at rest, it has some mass that we call the "rest mass."

If the object is moving, it has a certain amount of energy, called kinetic energy. The kinetic energy increases as the speed of the object increases. From the mass–energy equivalence relation, this kinetic energy is equivalent to certain amount of mass. Thus, the moving object has an effective mass that is the sum of its rest mass and the mass equivalent to its kinetic energy. The total mass of the moving object, as seen by an observer at rest, is therefore larger than the "rest mass." As the mass equivalent to the kinetic energy increases with the speed, the total mass increases with increasing speed of the object. Einstein showed that the mass of the object approaches infinity when it moves close to the speed of light.

15.7 Ultimate speed in the universe

An important consequence of the theory of relativity is that nothing can move faster than the speed of light which is 300,000 kilometers per second. This conclusion puts an upper limit on how fast we can travel, particularly, to explore the cosmos. As discussed before, the closest star from our planetary system is about four light years away. This means that a spacecraft, moving with the highest possible speed, the speed of light, cannot reach this nearest object outside our planetary system in any time less than four years. This makes it highly unlikely that humans will ever be able to visit the stars and galaxies that we see at night in the sky.

There are many ways to understand how this speed limit is imposed by the theory of relativity. We recall that, according to the first postulate of the theory of relativity, a person inside a spacecraft which is travelling at a constant speed in a single direction cannot determine whether his spacecraft is at rest or in a uniform motion. What we can show is that, if a spacecraft moves with a speed greater than the speed of light then it becomes possible for someone inside the spacecraft to determine whether the spacecraft is at rest or in uniform motion, thus violating the theory of relativity. The only way to preserve the theory of relativity is if nothing moves faster than the speed of light. Here we argue how this upper limit on how fast an object can move is obtained from the laws of relativity.

Let us suppose that a spacecraft is able to move with a speed greater than the speed of light. Let a person at the back of the spacecraft send a light signal toward the front end of the spacecraft. According to Einstein's second postulate of relativity, the light signal moves with the speed of light, 300,000 kilometers per second, no matter how fast the spacecraft is moving. This signal will therefore never reach the front end of the spacecraft as the speed of the spacecraft is larger than the speed of the light signal. The light signal cannot catch the spacecraft. The situation is similar to two runners who are running with constant speeds on a straight track in the same direction and are at the same location at some time. If the speed of the first runner is slower than the speed of the second runner, she will never catch or surpass the second runner. In our example, the light signal is like the first runner and the spacecraft like the second runner. The person inside the spacecraft will notice that the light signal never reaches the front. However, if the spacecraft was at rest, the light signal will hit the front in a time equal to the length of the spacecraft divided by the speed of light.

Now we have a contradiction. If the signal does not hit the front, the person inside the spacecraft concludes that the spacecraft is moving with a speed greater than the speed of light and if it hits the front, the spacecraft is at rest. This should not be possible according to the theory of relativity. The person inside the spacecraft who is moving at a uniform speed should not be able to distinguish between the state of motion and the state of rest. Therefore, the only conclusion we can draw, that is consistent with the theory of relativity, is that the spacecraft, and for that matter any object, cannot move faster than the speed of light. The speed 300,000 kilometers per second is the ultimate speed.

There are other ways of seeing this speed limit. We have seen in earlier sections that, when an object reaches the speed of light, the length shrinks to zero, clocks come to a standstill, and the mass becomes infinite for an observer on the ground. If the object could move with a speed faster than the speed of light, length, time, and mass lose their meaning. They are no longer real quantities. In the next section, we see that a communication faster than the speed of light can lead to a violation of causality, that is we can go to the past and change it affecting our present. This is truly a paradoxical situation. Speed of light has to be the ultimate limit.

It may be mentioned that speeds very close to the speed of light have been achieved by small particles like protons in particle accelerators. However, speeds larger than the speed of lighter are never seen in any accelerators in conformity with Einstein special theory of relativity.

15.8   Can we violate causality?

The speed of light being the ultimate limit can be justified on another basis. One of the tenets of physics is causality. This means that the present depends on what happened in the past up to the present time, but it does not depend upon what would happen in the future. Another way of defining causality is to say that every effect, like the movement of a rock, has a cause, like a push. The cause comes always before the effect. It never happens that the effect precedes the cause. If that happens then the entire universe will go haywire. What we show in the following is that if information can be transferred faster than the speed of light, then it should be possible to overcome causality.

Let us consider a simple example of two persons, Alice at rest on earth and Bob flying away in a rocket at a speed of 0.87 times the speed of light. For simplicity's sake, we assume that it is possible to communicate between Alice and Bob almost instantaneously. This means that the signals are exchanged between Alice and Bob without any time delay. This is the ultimate case of faster-than-light communication. The analysis become simpler for this extreme case and it helps to illustrate the proposition that faster-than-light communication can lead to non-causal effects and make the time-travel to the past possible.

We analyze the situation from both Alice's and Bob's points of view. From Alice's point of view, she is at rest and Bob is moving with a speed of 0.87 times the speed of light. However, from Bob's point of view, he is at rest inside the rocket and Alice is moving away in the opposite direction at the speed of 0.87 times the speed of light. Also, we note that Alice concludes that Bob's clocks are slower than hers. With Bob moving at 0.87 times the speed of light, his clocks are twice as slow as Alice's—Bob's clock ticks one second after every two seconds on Alice's clock. Bob, on

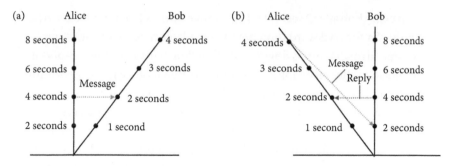

Fig. 15.6 (a) Alice is at rest and Bob is moving to the right at a speed of 0.87 times the speed of light. Alice sends a message after 4 seconds according to her clock. Bob's clock is slow according to Alice and he receives the message when his clock ticks 2 seconds. We assume instantaneous communication. (b) Same situation from Bob's point of view. Bob is at rest and Alice is moving to the left. Alice's message seems to be received by Bob even before it was sent. If Bob sends a reply after two seconds, it is received by Alice when her clock ticks 2 seconds. Alice gets the reply before she even sent the message.

the other hand, concludes just the opposite. For him, Alice's one second is equal to Bob's two seconds.

This situation is illustrated in Fig. 15.6. The space-time movements are depicted in Fig. 15.6a from Alice's point of view and, in Fig. 15.6b, from Bob's point of view. Here the movement is represented along the horizontal axis and time along the vertical axis. In Fig. 15.6a, Alice is at rest and there is no movement by her in the horizontal direction. However, Bob's rocket is moving farther away from Alice to the right as time passes. In Fig. 15.6b, Bob is at rest and Alice is moving to the left. Again, we should emphasize that both the figures are depicting the same situation but from two different perspectives.

Alice and Bob synchronize their clocks when they are at the same location. Now suppose that Alice sends a message to Bob after 4 seconds. Since we assume instantaneous communication, the message is received by Bob when his slow clock ticks 2 seconds. So far, everything appears normal. However, if the situation is looked at from Bob's point of view (Fig. 15.6b), Alice's message sent after 4 seconds according to her clock is received by Bob after 2 seconds. The strange thing is that time seems to be running in the backward direction, from a later time to an early time— the time arrow is in the downward direction. This is not all. Suppose, immediately after receiving Alice's message, Bob writes a response within two seconds and sends the reply to Alice. Again, assuming instantaneous communication, Alice receives Bob's reply after two seconds according to her clock. This is a truly paradoxical situation—Alice received the reply from Bob (after two seconds according to her clock) BEFORE she sent the message to Bob (4 seconds according to her clock). Causality is violated—the reply is received before the message was sent!!

This example shows that instantaneous communication between Alice and Bob violates causality. Indeed, it can be shown that the violation of causality is possible for any faster-than-light communication.

The theory of relativity therefore puts an upper limit on how fast an object can travel and how fast we can communicate with each other.

16

General theory of relativity

Space-time tells matter how to move; matter tells space-time how to curve.

—*John Wheeler*

The special theory of relativity was formulated by Albert Einstein in 1905. It describes how space and time are no longer independent of each other for two observers moving with respect to each other at a constant speed. As seen in the last chapter, the postulates of the theory of relativity led to unbelievable consequences: From the perspective of the person at rest, the clock runs slower and the length gets contracted for a person moving with a constant speed. Another important and consequential outcome is that energy and mass are interconvertible.

In 1907, while sitting in his office of the patent office in Bern where he had recently been promoted to Technical Assistant Second Class, Einstein had a thought that he would describe as the "the happiest thought of my life." Einstein imagined that a person falling from the roof of his house would feel weightless, as if floating freely in the space far away from earth with no pull of gravity. According to him "If a person falls freely, he will not feel his own weight." This simple thought would lead him to formulate the general theory of relativity that would become, along with quantum mechanics, one of the two pillars of modern physics. The theory would have far-reaching consequences in revolutionizing our understanding of space-time and would play a central role in comprehending the cosmos, its beginning and evolution, and possibly the end.

Einstein presented the basic ideas of general relativity in 1907. Central to this revolutionary theory was that space and time are curved in

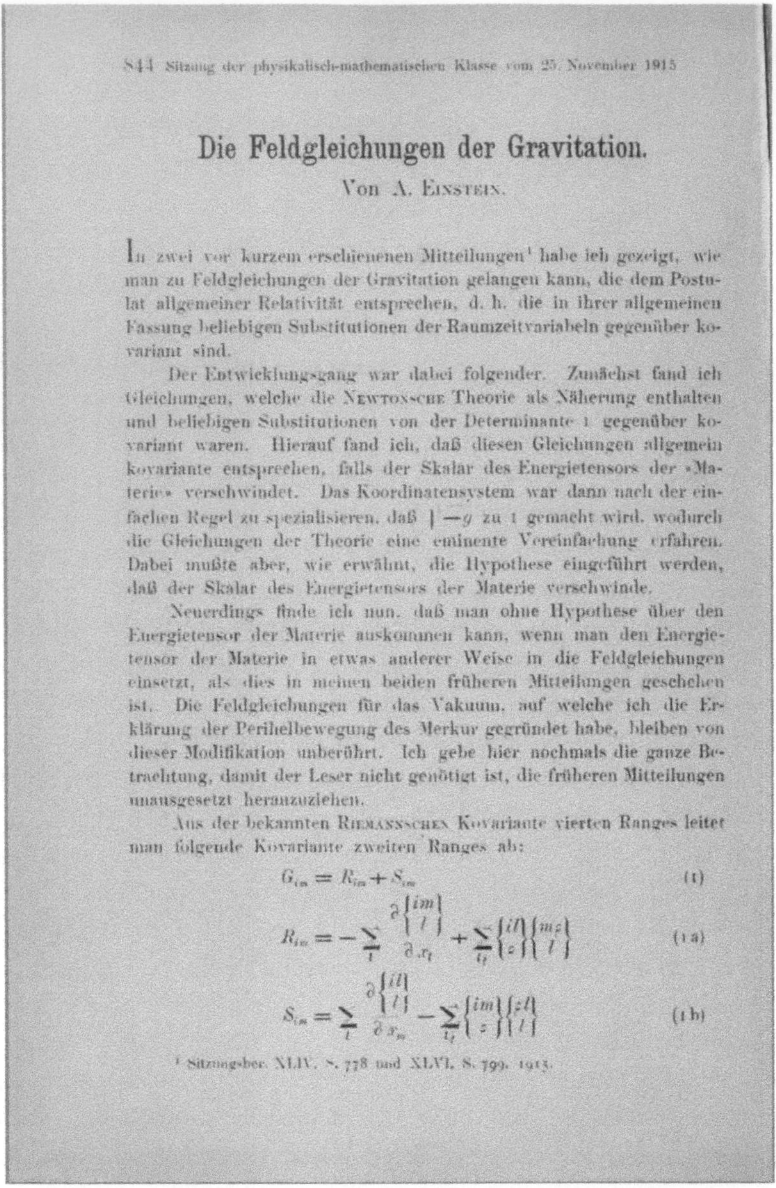

844 Sitzung der physikalisch-mathematischen Klasse vom 25. November 1915

Die Feldgleichungen der Gravitation.

Von A. Einstein.

In zwei vor kurzem erschienenen Mitteilungen[1] habe ich gezeigt, wie man zu Feldgleichungen der Gravitation gelangen kann, die dem Postulat allgemeiner Relativität entsprechen, d. h. die in ihrer allgemeinen Fassung beliebigen Substitutionen der Raumzeitvariabeln gegenüber kovariant sind.

Der Entwicklungsgang war dabei folgender. Zunächst fand ich Gleichungen, welche die Newtonsche Theorie als Näherung enthalten und beliebigen Substitutionen von der Determinante 1 gegenüber kovariant waren. Hierauf fand ich, daß diesen Gleichungen allgemein kovariante entsprechen, falls der Skalar des Energietensors der »Materie« verschwindet. Das Koordinatensystem war dann nach der einfachen Regel zu spezialisieren, daß $\sqrt{-g}$ zu 1 gemacht wird, wodurch die Gleichungen der Theorie eine eminente Vereinfachung erfahren. Dabei mußte aber, wie erwähnt, die Hypothese eingeführt werden, daß der Skalar des Energietensors der Materie verschwinde.

Neuerdings finde ich nun, daß man ohne Hypothese über den Energietensor der Materie auskommen kann, wenn man den Energietensor der Materie in etwas anderer Weise in die Feldgleichungen einsetzt, als dies in meinen beiden früheren Mitteilungen geschehen ist. Die Feldgleichungen für das Vakuum, auf welche ich die Erklärung der Perihelbewegung des Merkur gegründet habe, bleiben von dieser Modifikation unberührt. Ich gebe hier nochmals die ganze Betrachtung, damit der Leser nicht genötigt ist, die früheren Mitteilungen unausgesetzt heranzuziehen.

Aus der bekannten Riemannschen Kovariante vierten Ranges leitet man folgende Kovariante zweiten Ranges ab:

$$G_{im} = R_{im} + S_{im} \qquad (1)$$

$$R_{im} = -\sum_l \frac{\partial \begin{Bmatrix} im \\ l \end{Bmatrix}}{\partial x_l} + \sum_{l\rho} \begin{Bmatrix} i\rho \\ \rho \end{Bmatrix}\begin{Bmatrix} m\rho \\ l \end{Bmatrix} \qquad (1\,a)$$

$$S_{im} = \sum_l \frac{\partial \begin{Bmatrix} il \\ l \end{Bmatrix}}{\partial x_m} - \sum_{l\rho} \begin{Bmatrix} im \\ \rho \end{Bmatrix}\begin{Bmatrix} \rho l \\ l \end{Bmatrix} \qquad (1\,b)$$

[1] Sitzungsber. XLIV, S. 778 und XLVI, S. 799, 1915.

Fig. 16.1 Einstein's general relativity equations were first published on November 25, 1915 in German in the Proceedings of the Royal Prussian Academy of Science. The title translates to: "The field equations of gravitation."

the neighborhood of a massive object. In such a curved space-time man-ifold, the objects follow trajectories that may not be straight lines. For example, earth orbits the sun because space-time is curved in the vicin-ity of the sun. This is similar to the path of a ball on the surface of a trampoline with someone standing in the middle. This picture is very different from Newton's description of planetary motions as a result of the force of gravitation. Einstein described gravitation not as a force but a consequence of motion in a curved space-time geometry.

The physical ideas for such a curved space-time structure were quite simple. However, the mathematical formulation was a difficult task and relied upon the geometry of curved structures. This was a subject that had been studied in the nineteenth century by several mathematicians with a feeling that this may never find an application. It took Einstein sev-eral years to master these mathematical tools and derive the mathemat-ical equations that described how, in the presence of mass and energy, the structure of space and time evolves. These equations, first presented in December 1915, soon became a playground to explain certain unre-solved observations and predicted new, often amazing, consequences (Fig. 16.1).

Just as the special theory of relativity is based on the postulate that the speed of light is constant, the general theory of relativity is based on the equivalence principle under which the states of accelerated motion and being at rest in a gravitational field are physically identical.

16.1 Equivalence principle

In order to understand the equivalence principle, let us go back to Ein-stein's happiest thought and consider someone falling from the top of the building on earth—an unfortunate scenario but a useful one to grasp the concept. According to the Newtonian law of gravitation, there is a force of attraction toward the earth which accelerates the fall of the person. The acceleration rate is about 10 meters per second per second. This means that the speed with which the person is falling increases by 10 meters per second every second. Of course, we do not consider what happens when this person hits the ground.

Let us first consider the situation from the perspective of the falling person. He is falling under the action of only one force, the force of gravity. The important point is that, he experiences weightlessness. If he is standing on a weighing machine, the weighing machine is also in the free fall. His weight will be zero as no force is exerted on the weighing machine. Similarly, if, during the fall, he releases an object like a ball, both the falling person and the ball move with the same acceleration, the acceleration due to gravity. They will therefore stay stationary with respect to each other. The falling person will see the ball suspended in space.

Next let us consider a slightly different situation. Instead of being in free space near earth, he is inside a compartment which is freely falling as shown in Fig. 16.2a. What will this person observe? Again, the person will feel weightlessness. If he stands on the weighing machine inside the closed freely falling compartment, his weight will be zero. In addition, he would see all the objects around him inside the compartment to be static and not moving.

This feeling of weightlessness is identical to the feeling of an astronaut in space, far away from earth as shown in Fig. 16.2b. The astronaut also sees the objects around him to be static and at rest, just like the person inside the freely falling compartment on earth.

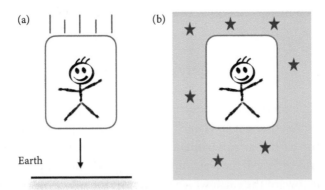

Fig. 16.2 Einstein's happiest thought: A person in an elevator falling freely on earth would feel the same weightlessness as being in empty space away from any gravitational pull.

Thus, a person inside the compartment who is unable to look out will not be able to distinguish between the two possible situations: when the compartment is falling freely on earth and when the compartment is static and freely floating far away from the earth.

Next let us consider the same person again in a closed compartment. We again consider two situations. First the compartment is resting stationary on earth (Fig. 16.3a). Imagine that there is a weighing machine in the compartment. When this person steps on the weighing machine, he finds his weight to be (say) 80 kilograms. The weight is essentially the force that the person exerts on the weighing machine.

Next, suppose that this compartment is taken to space far from the earth where there is no force of gravity. The weighing machine records zero weight as discussed above. However, if the compartment is accelerated upward, as shown in Fig. 16.3b, with an acceleration of 10 meter per second per second, the weighing machine is pushed in the downward direction with a force according to Newton's laws of motion. He will again weigh 80 kilograms.

The result is that the person will not be able to distinguish between the force of gravity and the Newtonian force on the weighing machine

Fig. 16.3 Equivalence principle: A person doing any experiment, like dropping an object, inside a container on earth should see the same outcome as in a container moving an acceleration equal to the acceleration due to gravity in the upward direction.

due to upward acceleration as, in both cases, the weighing machine will record the same weight, 80 kilograms. For him the two situations, being stationary on earth or accelerating upward in outer space, are identical. The only way to find out for him whether he is on earth or in space will be to peek out of the closed compartment.

On the basis of this simple argument, Einstein formulated an equivalence principle which can be stated as follows: The states of the accelerated motion and being at rest in a gravitational field are physically identical.

The foundation of the general theory of relativity and Einstein's theory of gravitation were laid on this simple principle as we see in the following.

16.2 Curved space and time

A question of interest is: Is there a way to distinguish between the two situations, the states of the accelerated motion and being at rest in a gravitational field? If we could do that, the equivalence principle will no longer be true. Here we assume that the size of the compartment is small so that the gravitational pull toward the earth is uniform throughout the compartment.

A possible experiment is to take a flash light and point it horizontally on the wall. If the compartment is accelerating upward, the light beam will bend downward as shown in Fig. 16.4. But why? When the light beam leaves the flashlight, it moves horizontally at the speed of light. However, by the time it reaches the wall, the wall has moved up as the compartment is accelerating upward at 10 meters per second per second. Thus, light hits the wall below the level it was sent, and the light appears to bend downward (See Fig. 16.4d). This result is straightforward and even Newton would have agreed.

But what happens if the compartment is at rest on the surface of earth as shown in Fig. 16.5b? Will the light beam get deflected as in the accelerated compartment as shown in Fig. 16.5a? Common sense tells us that the light will not bend and go straight and will hit the wall at A in Fig. 16.5b which is at the same height as the flashlight. The only situation in which the light will bend is if light consists of particles that

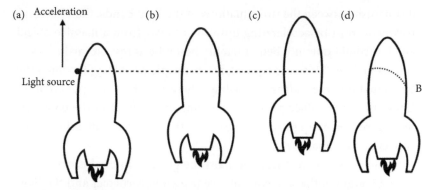

Fig. 16.4 (a–c) Light pulse is sent in the horizontal direction in a rocket moving in the upward direction with an acceleration equal to the acceleration due to gravity will be moving in the horizontal direction while the rocket moves upward. (d) However, inside the rocket, the light pulse will appear to bend in the downward direction and hit the other end below the horizontal.

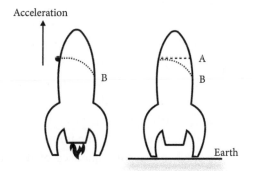

Fig. 16.5 If the light pulse is sent in the rocket that is accelerating in the upward direction, the light bends. For a rocket standing still on earth, the light pulse should travel in the horizontal direction and hit the other side at point A. However, if the equivalence principle is correct, it should bend and hit at point B.

are massive and can be deflected due to a force of gravity. But we know that light has no mass, therefore no deflection is expected.

If this argument is correct, then the person inside the compartment, who is unable to see outside, can find out whether he is at rest on earth surface or accelerating in the far-away space. He should be able to distinguish between the two situations: If the light bends, then his compartment must be accelerating upward far away from a massive object and if the light does not bend then he should be at rest on earth.

This violates Einstein's equivalence principle according to which the two situations are indistinguishable. There are thus two possibilities: Either the equivalence principle is wrong or the light in the compartment at rest on earth should also bend by the same amount as inside the accelerated compartment.

Einstein took a great leap, perhaps the greatest in the history of science, by arguing that the equivalence principle is correct and the light indeed bends on the earth-bound compartment by the same amount as in the compartment moving upward with the same acceleration as the acceleration due to gravity.

But how can that be? Light always takes the shortest path to go from one point to another. It never zigzags to go from one point to another. As a result, light should go straight in the compartment that is at rest on earth and hit at the same height as the flashlight. Einstein postulated that indeed light moves across the compartment along the straight and shortest path. However, the bending is caused because space becomes curved in the presence of earth and the light takes the shortest path, not in a flat space, but in a curved space. This was an amazing insight—space that was considered absolute and flat indeed curves in the neighborhood of massive objects like the earth, sun, planets, and stars. This was a giant step that changed our concept of space-time forever.

Before we discuss further the consequences of this mind-boggling idea, let us first understand how the bending of space can give rise to the bending of light inside the compartment. A simple example, as shown in Fig. 16.6, can illustrate that. If the space is flat, as we perceive it around ourselves, the light follows a straight path which is the shortest path to go from point A to B. But what happens if the space is curved like the surface of the sphere? The surface of the sphere is a two-dimensional space that is curved and is definitely not flat. If the light is constrained to move on the two-dimensional surface of the sphere, it will have to bend in order to follow the shortest path to go from point A to point B. This explains the bending of light in a curved space. More about the properties of the curved space is presented in the next section.

(a)

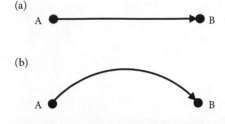

(b)

Fig. 16.6 The distance between two points in a flat space is shorter than the same two points in a curved space.

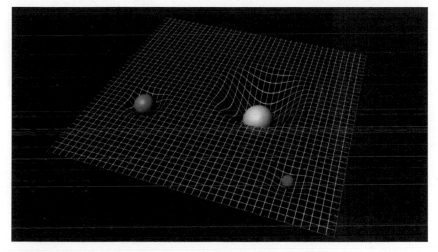

Fig. 16.7 A pictorial way of seeing how massive objects can warp space-time. Here space-time is represented by a two-dimensional surface which is distorted by three massive bodies. The distortion is proportional to the mass of the object.

(Credit: ESA- C. Carreau)

Einstein's theory of gravitation was a dramatic departure from Newton's law of gravitation. According to Newton's law, an object falls on earth because there is a force of gravitation between the object and the earth. Now Einstein was telling us that the reason the object falls on earth is due to the curvature or bending of space in the presence of earth. Gravity, in Einstein's theory of relativity, was no longer described as a force. Rather it was described in terms of the geometry of the space. It is due to the bending of space that the object is made to move in the prescribed path.

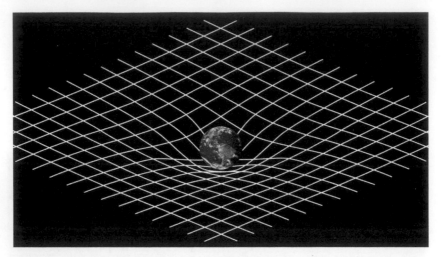

Fig. 16.8 Massive objects cause space-time to curve, much like a heavy ball will create a well in a stretched-out piece of fabric.
(Credit: NASA)

This is the kind of stuff that science fiction is usually made up of. However, here the universe is being described in these unbelievable terms.

Next, what about time? How does gravity affect time? According to the theory of relativity, space is curved near the surface of a massive body like earth as compared to distances far away where the space is flat. A consequence is that time runs slower where the gravity is the strongest. But why?

We can argue the "gravitational time dilation" as follows. Consider a clock consisting of two points A and B such that light is sent from point A to point B. A unit time is defined as the time taken by the light to go between the two points. As mentioned above, light takes the shortest path to travel between the two points. Far away from earth, space is flat and light takes the time equal to the distance between A and B divided by the speed of light. Now, according to Einstein, the speed of light is a constant and does not depend on where we are. Also, space bends near a massive object. Thus, if the same two points A and B are close to earth, the shortest distance on the curved path will be longer than in the flat space away from the massive body. Therefore, as the speed of light remains the same, the time travelled from A to B is longer. Thus, one second on the surface of earth is longer than the one second recorded in space.

Here we have considered a very specific type of clock. An obvious question is whether our watches and the usual clocks will show similar behavior. The answer is yes! We can synchronize our watches and clocks to the clock discussed above. All of them should show the same time.

A consequence of our new vision of space-time is that we can expect to live longer on a more massive planet like Jupiter than on earth. Here the difference may be too small—we may live only few minutes longer on Jupiter than on earth. However, if we can travel to a much more massive object like a neutron star, our life span could be much longer. Near a black hole, we can, in principle, live forever.

The universe is truly mysterious and strange if all this is true.

But how to prove that space-time is curved? There have been several tests over the last hundred years. Here, we mention some of the classic tests that established the correctness of this incomprehensible nature of the universe.

Before discussing these tests of Einstein's theory of relativity, it should be noted that the gravitational time dilation has consequences for the Global Positioning System (GPS). The GPS satellites are positioned about 12,550 miles above Earth's surface, far away from the earth's gravitational field. The clocks on these satellites tick faster than the clocks on earth's surface. Therefore, their internal clocks are adjusted in order to take into account their faster time so that the GPS data sent back to Earth's surface have matching times. Without this correction, GPS satellites would not be the useful tool that we know them to be.

Time dilation due to the space-time warping due to gravity can also lead to science fiction scenarios. Suppose someone 20 years old in the year 2022 would like to see how the earth would look like two hundred years later, in the year 2222. One possibility is that she lives a long life of 220 years. In the present scenario, this seems highly unlikely as the longest living humans live for not much more than 100 years. However, space-time warping offers another possibility.

She can travel from earth to a nearby black hole where the gravity is so strong that the clocks, and time, run so slowly that 40 years near its surface are equivalent to 180 years on earth. If it takes 10 years to travel from earth to the black hole on a superfast rocket, she can spend 40 years in the neighborhood of the black hole and then return to earth on the same superfast rocket. When she returns, her age is 80 years and she would expect to be back on earth in the year 2102. However, at earth the year will be 2222.

Of course, there are many technical issues for this scenario to be realistic. We do not know of any black hole that exists so close to earth. And it does not appear realistic to survive in the vicinity of the black hole. Also, we ignored the time dilation effects due to the superfast travel to the black hole. Nevertheless, Einstein's general theory of relativity seems to open the possibility for such counterintuitive scenarios.

16.3   Why does an object fall on earth?

Let us examine how the theory of relativity can explain the motion of objects under gravitation. In Newtonian mechanics, an object moves in a certain trajectory as a result of the gravitational force. If we throw an object upward with a certain speed, it will rise, slow down due to the force of gravity, stop mid-air, and then fall on the ground. All this can be explained by Newton's law of gravitation and the laws of mechanics. In a similar manner, the motion of the moon around the earth and the motion of the planets around the sun can be understood by the same laws.

According to Einstein's theory, there is no "force." The reason why an object moves along a certain trajectory is because it is moving in a straight line, not in a flat space-time but in a curved space-time. Massive objects like the earth and the sun are responsible for the curvature or warping of space and time. Let us illustrate with an example.

Consider a ball is thrown in a horizontal direction on earth. It follows the trajectory as shown in Fig. 16.9. The ball, while moving in the horizontal direction is also falling down, eventually hitting the ground. According to Newtonian mechanics, the downward vertical motion is

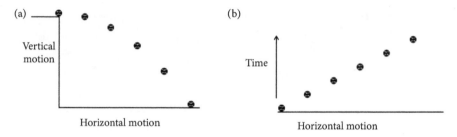

Fig. 16.9 A ball thrown in the horizontal direction follows the trajectory as shown in (a). There are two components of motion, horizontal and vertical. The ball moves in the horizontal direction with the same speed with which it is thrown. However, the vertical motion is determined by the acceleration due to gravity. In (b) the horizontal motion is plotted as a function of time. The straight line is a sign that the ball moves with constant speed in the horizontal direction.

due to the force of gravity leading to a curved path. However, according to Einstein, the ball does not feel a force of gravity. Instead, it is moving freely within the tapestry of space and time. What we learn from Einstein's theory is that the path of the ball is a straight line in four-dimensional space-time (three-dimensional space plus one-dimensional time). The technical name for this path is "geodesic." This picture is quite simple, but hard to imagine. As a special case, if there is no curvature in space-time, such as in regions far away from a massive object, we see an object either at rest or moving with a uniform speed in the flat space.

A question immediately comes to mind. If space-time is curved, then how come two objects thrown horizontally with different speeds end up at different locations on the ground as shown in Fig. 16.10a. The curvature seen by both objects is the same. Therefore, we would expect the same straight path (in curved space-time) for both of them. Both of them should therefore land on the floor at the same location. But they don't.

An explanation is based on the curvature of not just the space but also the time. If only space was curved, then both objects would have landed at the same location. However, the theory of relativity has made us realize that space and time cannot be treated independently. They are

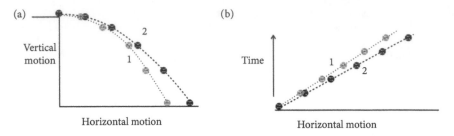

Fig. 16.10 (a) The trajectory of the two balls thrown with different speeds in the horizontal direction. At each instant, their horizontal and vertical locations can be determined. (b) The same motion where the horizontal motion is depicted at different times. If the motion of the two objects is looked upon as the motion in a straight line in the curved space and time, the warped space and time will be rather complicated.

curved or warped together. In Fig. 16.10b, we consider the horizontal distance travelled by the two objects at different times. What this figure shows is that, at a given time, the two objects have moved different distances in the horizontal direction. The two figures (Figs. 16.10a and 16.10b) are quite similar. There is, however, one big difference. In Fig. 16.10a, the trajectory of both objects is shown in the two-dimensional space corresponding to horizontal and vertical motion. However, in Fig. 16.10b, the trajectory is described in the mixed space and time regions. The left-right motion depicts the horizontal part of the motion and the corresponding time is depicted in the vertical direction.

If space and time constitute the four dimensions, the complete picture for the motion of the two objects should require a four-dimensional picture, which is hard to visualize. However, it is clear now that the motion of the two objects starts in the same direction in space, but they move in different directions in space AND time. They, therefore, see different curvature in the four-dimensional space-time domain. Consequently, the straight lines in the curved space-time domain are different for both objects and they end up at different locations.

This discussion clearly illustrates why all the events should be analyzed in the space-time domain and not just the space domain.

16.4 Embedding diagrams

According to Einstein's general theory of relativity, space and time are curved in the presence of massive objects. An object then follows a straight-line trajectory in this curved space-time. This straight path is called a geodesic. A geodesic is therefore a straight path in a curved space-time that appears to be a curved path in flat space and time. It is hard to visualize the space-time curvature, particularly for a three-dimensional space and one-dimensional time. Here we present a simple method via what are called embedding diagrams.

First let us consider two-dimensional surfaces embedded in three dimensions. Suppose creatures living on these surfaces can perceive only two-dimensional surfaces and cannot see the third dimension. The simplest example is a flat surface as shown in Fig. 16.11a. How can a creature living on this surface convince herself of the flatness of her universe? There are several ways of doing that, all of them requiring the basic laws of geometry. If she draws two parallel lines, they keep the same distance from each other and never meet, no matter how long these lines are. Another is that, if she draws a triangle as shown in Fig. 16.11a, the three inside angles add to 180 degrees. Finally, if she draws a circle, the circumference is equal to a constant called pi (approximately equal to 3.1417) times the diameter of the circle. So, without being aware that her two-dimensional universe is embedded in a third dimension, she can confirm the flatness of her universe.

(a)   (b)

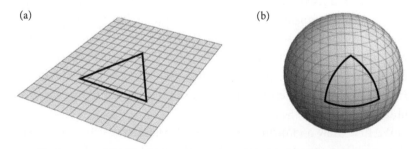

Fig. 16.11 (a) Flat surface, (b) spherical surface.

Next suppose, her universe is curved. A simple example of a two-dimensional curved surface is the surface of a sphere (Fig. 16.11b). How can a creature confined to this two-dimensional space find whether her universe is curved and by how much? She can again draw a triangle. This time the sum of the angles exceeds 180 degrees. Another way is to draw a circle and measure both the circumference and the diameter. In this case, the circumference will be smaller than pi times the diameter. The smaller the ratio of the circumference and the diameter, the more curved is the surface. We note that the circumference is equal to pi times the diameter as observed by the observer in the three-dimensional world. The plane of the circle is inside the sphere. However, the creature living in the two-dimensional universe can only see the surface and not the interior of the sphere. She may not even know that her universe is the surface of a sphere. Thus, when she moves in a certain direction on the surface of the sphere, she comes back to her original location and the distance travelled becomes the circumference. In order to determine the diameter, she travels the shortest path between the starting point and the farthest point on the circle. The diameter is larger than the circumference divided by pi. The ratio of the circumference and the diameter is a measure of the amount of curvature.

An important point to note is that, to get a good understanding of the curvature of a two-dimensional surface, we visualized it embedded in a three-dimensional space. The third dimension is not real for the creature that is confined to only the two-dimensional surface. This same idea can be used to visualize the space-time curvature of a two-dimensional surface due to gravity in a curved three or four-dimensional space-time.

For simplicity's sake, we consider stars that are like spheres with uniform density. In Fig. 16.12, we consider three such stars with equal mass but different radii. According to the general theory of relativity, they create the curvature in space-time; the curvature is larger for stars with smaller radii (larger density). For such stars, the embedding diagram is obtained as follows.

We first draw a two-dimensional equatorial plane crossing the star. Far away from the star, the surface is flat where the laws of conventional geometry are valid: parallel lines remain parallel and the circumference of a circle is pi times the diameter. However, near and inside the

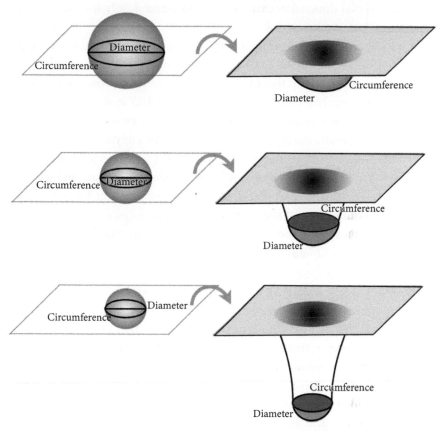

Fig. 16.12 Embedding diagrams showing the curvature of a two-dimensional space in three dimensions. The curvatures of the equatorial planes of the spherical objects of equal mass but decreasing radius are shown on the right. The curvature of the two-dimensional plane in the presence of massive object is shown in three-dimensions. The third dimension not real.

star, the space is curved and if we visualize the two-dimensional surface of the equatorial plane in a three-dimensional surface as above. The surface bends as shown in Fig. 16.12. The circumference of the circle is smaller than pi times the diameter in this region. A higher curvature corresponding to a higher density star is shown by a deeper dip in the three-dimensional space. The third dimension in these figures is an external dimension that we have introduced only for the sake of convenience and is not real.

Such embedding diagrams are very useful in analyzing and visualizing the space-time curvature. Thus, the most common picture depicting the space-time curvature is that of a rubber sheet weighed down by a heavy spherical object like a marble as shown in Figs 16.7 and 16.8. It should, however, be kept in mind that these figures only roughly approximate the actual curvature arising from Einstein's theory of relativity.

There is another problem. The embedding diagrams of the type depicted in Figs 16.7 and 16.8 show only the bending of space and ignore the time curvature. A complete picture, as discussed in Section 16.3, involves both space and time curvature. For many cases including planetary motion around the sun, the curvature of time may have a greater effect than the curvature of space.

16.5   Bending of light: Newton vs Einstein

Einstein published his paper, in which he described gravity as a geometric property of space and time, in 1915. There were attempts to see whether these revolutionary ideas about space and time were correct. This motivated one of the most dramatic experiments in the history of physics which would pit the two giants of science, Isaac Newton and Albert Einstein, and their conflicting theories of gravitation against each other. The experiment was the bending of light by a massive object.

We recall that Newton championed a corpuscular nature for light. Newton had also noted, while formulating his theory of gravitation, that any material particle moving at a finite speed would experience a force while passing in the vicinity of a massive object. This pull by gravity should bend the trajectory of the particle and the bending angle should be independent of the mass of the particle. Thus, if light is composed of small particles, they should also experience such deflection. Newton himself did not calculate this deflection as, in his time, the finite speed of light was not well established. However, he postulated this deflection and, toward the end of his treatise *Opticks* (1704), he noted, "Do not Bodies act upon Light at a distance, and by their action bend its Rays, and is not this action strongest at the least distance?" The finite speed of light was well established by the early nineteenth century and, in 1804, a German astronomer, Johann Georg von Soldner, presented calculations based on Newton's corpuscular theory that light weighs and bends like high-speed projectiles in a gravitational field. He produced a value of 0.87 arc seconds bending angle for light grazing the sun.

In 1911, more than a hundred years later, Einstein calculated the bending of light by combining the equivalence principle with the special theory of relativity to predict a deflection of light from the sun by the angle of 0.87 arc seconds. This is the same value that Newtonian theory predicted. He obtained this result before he formulated the general theory of relativity and the associated curved space-time. When he included the effects of the general theory of relativity, the predicted value for the bending of light doubled to 1.83 arc seconds. This result was published on November 18, 1916.

Thus, the predictions of Newton and Einstein were at odds with each other and an experimental activity followed soon to decide who was right. The bending of light by a massive object also became the first test of the esoteric general theory of relativity.

It is impossible to see the bending of a light ray coming from a distant star by the sun due to the brightness of the sun (Fig. 16.13). The only possibility is to see the bending of light during a solar eclipse when the moon blocks the sun. If the solar eclipse is complete, then the grazing light from the edge can provide information about the bending angle of the light. This is a difficult measurement.

After the First World War was over, Arthur Eddington organized an expedition to the island of Príncipe off the coast of Africa to watch the solar eclipse on May 29, 1919 and to measure the observed curving of light from distant stars by the gravitational pull of the sun. While the expedition was being planned, Eddington wrote: "The present eclipse expeditions may for the first time demonstrate the weight of light (i.e., Newton's value) and they may also confirm the added effect of Einstein's weird theory of non-Euclidean space, or they may lead to a result of yet more far-reaching consequences of no deflection."

When the results were announced, they agreed with Einstein's predicted value. Einstein overnight became an international celebrity and an iconic figure.

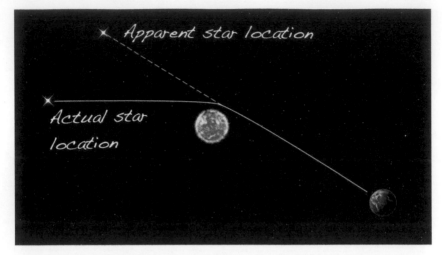

Fig. 16.13 Bending of light by a massive object like the sun. Light coming from a star passing near the sun is deflected slightly by the "warping" of space-time. We assume the light beam has been traveling in a straight path throughout its journey. The position of the star appears to be shifted from its true position. If we were to observe the star at another time, when the sun is not in the way, we would measure its true position.

(Credit: NASA)

16.6 Gravitational lens

The bending of light by a massive object can lead to another phenomenon, the gravitational lensing effect, that has emerged as an important measurement tool in cosmology.

In order to illustrate this effect, let us first consider the focusing effect of a lens. A lens is a piece of glass or other transparent material which can bend the path of light as shown in Fig. 16.14a. The basic principle behind the focusing effect is that the speed of light is slower in the glass as compared to that in air. The consequence is that a wavefront travels much faster at the edges where it has to travel a shorter path inside the lens than in the center where it has to travel a longer path. The result is that the wavefront changes direction after the passage of the lens and becomes focused at a point.

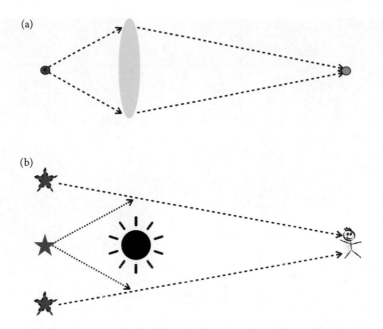

Fig. 16.14 (a) Optical lens: Light from a point object is bent by a lens. The lensing effect is due to the reduced speed of light inside the glass as compared to the speed of light in air. (b) The gravitational lensing effect from an object like a star is due to the bending of light by a massive object like the sun. If observed from earth, the light from star will appear to come from a displaced location.

In the gravitational lens, a massive object like a galaxy acts like a lens (Fig. 16.14b). As seen in the last section, light bends in the presence of a massive object. This happens because space and time curve in the neighborhood of the massive object whereas light propagates in a straight line.

We consider a bright object like a supernova beyond a galaxy. Let us assume that the galaxy is made up of homogeneous mass. This seems like a strange assumption knowing that a galaxy consists of billions of stars but that each star is typically several light years (trillions of miles) away from its neighboring stars. How can such a sparse galaxy be considered a homogeneous distribution of mass? The situation is

Fig. 16.15 Image taken by Hubble Telescope on August 27, 2021. The two bright lights inside the ring are two galaxies that act as a gravitational lens in space. The light from a distant quasar behind the galaxies has been bent by the curved space around the galaxy pair. The ring and the four points come from light originating from the quasar.

(Credit: ESA/Hubble & NASA, T. Treu. Acknowledgment: J. Schmidt)

similar to a speck of dust which consists of trillions of individual atoms. On a distance scale much larger than the distance between individual atoms, the dust particle appears completely homogeneous.

The light originating from the bright object on the other side of the galaxy is bent by the galaxy due to the space-time curvature and would appear to come from a different location if seen from earth. In the symmetric case when the galaxy is assumed to be spherical and the star is just on the opposite side, we, on earth, would see the light source (star) as a ring around the lensing object, usually called an Einstein ring. This is seen in Fig. 16.15. The size of the ring depends upon the gravitational bending by the galaxy, which in turn depends upon the total mass of the galaxy.

For different shapes and mass distributions of the galaxy, the bending of light would be different. Thus, gravitational lensing becomes a tool to study the properties of an entire galaxy such as its mass distribution.

16.7 Precession of the perihelion of Mercury

A long-standing problem since Newton's time was to account for the motion of the planet Mercury around the sun. Newton's law of gravitation and the laws of motion predicted an elliptical motion with the sun located on one of the focuses. Newton's theory predicted the ellipse to be fixed if there was no other planet present. However, it was observed that the perihelion (the point in the orbit at which it is closest to the sun) of Mercury was not fixed, but orbited around the sun as shown in Fig. 16.16. Thus, the whole elliptical orbit rotates around the sun as seen in the figure. As seen from the earth, the precession of Mercury's orbit is measured to be 5,600 seconds of arc per century (one second of arc is equal to 1/3,600 degrees). This is a small, but measurable, effect.

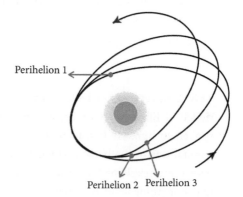

Fig. 16.16 The elliptical orbit of the planet Mercury around the sun. The point closest to the sun, called perihelion, undergoes rotation.

Perihelion 1

Perihelion 2   Perihelion 3

One way to account for this precession was to include the motion of other planets like Venus and earth in the calculations. The gravitational pull of these and other planets could affect the path that the planet Mercury followed. The mathematical problem that includes these extra factors makes the calculation quite complicated. However, it can be solved and it was found that Newton's laws could account for much of this precession. The new results predicted the precession of Mercury to be 5,557 seconds of arc per century. There is still a discrepancy of 5600 − 5557 = 43 seconds of arc per century.

How large is this? The distance of the perihelion of Mercury from the sun is 46 million kilometers. A shift of 43 seconds of arc is therefore approximately equal to a displacement of 9,600 km. This comes to 96 km per year. Another way is to construct an isosceles triangle with equal sides of 46 million kilometers and a base of 9,600 kilometers. Then the angle subtended will be 43 seconds of arc.

Attempts were made to explain this discrepancy. For example, it was postulated that just enough of an amount of dust exists between Venus and the sun that would explain this difference using Newtonian laws. However, no dust was found. This is where the situation stood when Einstein proposed his general theory of relativity in 1915.

According to Einstein, space-time is curved in the neighborhood of the sun. This curvature contributes further precession of Mercury around the sun. When Einstein calculated this precession using his theory, he got precisely the previously unexplained 43 seconds of arc per century. This was stunning evidence in support of Einstein's theory. This extremely small deviation that had remained unexplained for centuries could be explained via the novel concepts of space-time curvature. In the beautiful biography of Einstein "Subtle is the Lord…," Abraham Pais writes,

> This discovery was, I believe, by far the strongest emotional experience in Einstein's scientific life, perhaps in all his life. Nature had spoken to him. He had to be right. … his discovery had given him palpitations of the heart… when he saw that his calculations agreed with the unexplained astronomical observations, he had the feeling that something actually snapped in him.

Einstein's theory also correctly accounted for a smaller discrepancy of 8.6 seconds of arc per century in the precession of the perihelion of Venus. The value is smaller than that of Mercury because Venus is at a larger distance from the sun and the curvature of space-time is less.

16.8 Gravitational red shift

The third classic test of Einstein's general theory of relativity is the so-called gravitational red shift. The basic idea is that if a light pulse of high frequency is sent from a region of higher force of gravity, like the ground, to a region with lower force of gravity, like the top of a tall building, the frequency of light will decrease—blue light will tend to turn red. This is shown in Fig. 16.17. This result is a direct consequence of the curved space-time in the vicinity of the massive object, earth.

How do we understand this? As noted above, time moves slower near earth than farther away. For the sake of the argument, let us assume that a single second at the surface of earth is equal to half the second as measured at the top of the building. Also suppose a signal of frequency equal to 10 (10 crests per second) is sent from the surface of the earth. Then, during the diminished one second on the top of the building (equal to half a second on the surface of earth), only 5 crests corresponding to a frequency of 5 cycles per second will be received. The frequency has decreased for a signal to go from the surface of the earth to the top of the building.

In the real situations, the effect is extremely small. For blue light, the frequency shift on earth is much less than a trillionth of a wave. The effect

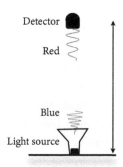

Fig. 16.17 Gravitational red shift. The wavelength of the light emitted on the surface of the earth increases as it propagates upward.

is, however, very large near objects with a much higher force of gravity. For example, the acceleration due to gravity at the surface of a neutron star can be 200 million times larger than that of earth (about two trillion meters per second per second). The gravitational red shift will be much more dramatic there.

16.9 Gravitational waves

One of the major predictions of Einstein's general theory of relativity was the existence of gravitational waves that are ripples in space-time created by certain cosmic events in the universe. Such events include the collision of massive objects like black holes and neutron stars and the exploding massive stars producing supernovae. The production of space-time ripples is similar to dropping a pebble in the lake producing water waves that move away. As long as a massive object like a black hole or neutron star remains undisturbed, the space-time structure does not change much. However, any cosmic event that involves violent changes in the space-time structures sends out gravitational waves. Gravitational waves travel at the speed of light and they can sqeeze and stretch anything in their path.

Gravitational waves were predicted in 1916. It took almost sixty years to collect indirect evidence for the existence of gravitational waves. In 1974, Joseph Taylor and Russell Hulse observed the motion of a pulsar around a neutron star about 21,000 light years away. This was the first time that a pulsar had been observed in a binary system. According to Einstein's general theory of relativity, such a binary system should generate strong gravitational waves. However, gravitational waves carry energy and this energy should come from the motion of the pulsar. This should cause the pulsar to lose energy, causing its orbit to contract. And this is exactly what was observed, justifying Einstein's prediction of gravitational waves.

It took another forty years until, in a landmark experiment, an international team of researchers spread over several continents detected gravitational waves directly. These gravitational waves were generated by two colliding black holes 1.3 billion light years away. The signal received on earth was squeezing and stretching by an amount much smaller than the size of a proton. It is mind-boggling to see directly the effect of an event that took place well over one billion light years (with one light year being equal to about 9.6 trillion kilometers) away. This experiment represents one of the greatest scientific achievements of all time.

The experiment to detect gravitational waves was based on a Michelson interferometer (Fig. 16.18). Michelson, as discussed in Section 15.2, had invented the interferometer in 1889 to detect the presence of ether. Almost 130 years later, essentially the same set-up was used in another landmark experiment, to detect gravitational waves. However, the size and complexity of these gravitational wave detectors were quite different from the first interferometer. In order to gain sensitivity, the two arms of the interferometer were about four kilometers in length. This is in contrast to the Michelson–Morley interferometer that was only 1.3 meters long. When a gravitational wave hits the interferometer, it changes the path length in one arm with respect to the other. This path difference is then detected by the movement of the fringes as discussed in Section 15.2. The amazing thing is that, for the gravitational waves

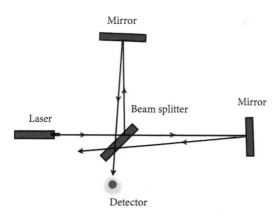

Fig. 16.18 Schematics for the Michaelson interferometer to detect the gravitational waves.

generated by cosmic events like the collision of two black holes trillions of miles away, the extremely weak signal produces a difference in path lengths about ten thousand times smaller than the size of a proton. This is an incredibly small distance.

17

Black holes

Why are black holes so different from all other objects in the macroscopic Universe? Why are they, and they alone, so elegantly simple? If I knew the answer, it would probably tell me something very deep about the nature of physical laws. But I don't know.

—Kip S. Thorne

The existence of black holes is among the most stunning consequences of Einstein's theory of general relativity. These are point-like massive objects whose gravitational pull is so strong that nothing can escape and time slows to a complete stop. This exotic object is the last stage in the death of a massive star.

The history of black holes goes back to the late eighteenth century when a British scientist, John Michell, and the famous French mathematician Pierre-Simon Laplace independently predicted, on the basis of Newtonian mechanics, the existence of stellar objects whose gravitational field could be so strong that nothing, not even light, could escape from their surface. Their ideas were based on the concept of "escape velocity."

It is an everyday observation that when a ball is thrown upward with a certain speed, it moves upward with a slowing speed, eventually stopping in the air at a certain height and then falling back to the ground. The reason that it slows down first and then falls back is gravity. The gravitational pull of the earth tends to slow down the ball. If the ball is thrown up with a higher speed, it will go a bit higher and then fall back on the ground.

A question of interest is: Is it possible for an object like a ball to be thrown at such a high speed that it goes up, up, and up and never comes back? The ball would then be free of earth's gravitation. It turns out that, on earth, it is possible for the ball to escape the earth's gravitation if it is thrown up with a speed of eleven kilometers per second or higher. This is called the "escape velocity." This speed is independent of the mass of the ball but depends on the ratio of the mass and the radius of the earth.

A practical demonstration is when spaceships sent to the moon and Mars should first break the shackles of gravity and then start the journey toward the stellar objects. Similarly, spaceships, sent to explore the solar system must first escape the earth's gravity. They are launched on earth with a speed larger than the escape velocity. The speed required to break the shackles of gravity is different on different stellar objects. For example, the escape velocity is about two kilometers per second on the moon, where the force of gravity is weak. This is why, during the Apollo landings on the moon, the lunar module did not need a large rocket engine to leave the moon and return to earth. The escape velocity on the sun is 620 kilometers per second.

From this discussion, it is clear that, for massive stellar objects with smaller radii, the gravitational pull is stronger, and consequently, the escape velocity is higher. For example, at a neutron star, the escape velocity ranges from 100,000 to 160,000 kilometers per second depending on the mass and the radius of the neutron star.

A question is whether it is possible to have an object whose gravitational pull is so strong that the escape velocity becomes equal to the largest speed possible according to Einstein's theory of relativity, the speed of light. If such an object exists, nothing, absolutely nothing, should be able to leave it. What about light? Would such an object allow light to escape?

In 1784, John Michell. and, in 1796, Pierre-Simon Laplace concluded that an object whose escape velocity is the speed of light would not allow light to escape. Therefore, when we would look at it, it would appear black, pitch black, because no light will be able to leave from such an object. This object was described as "a dark star." On this basis, Michell calculated that a star 500 times bigger than the sun, but with the same density, would have an escape velocity equal to the speed of light.

This is, however, not a correct picture. Implicit in this description is the assumption that light consists of particles with mass. This was the prevalent view of light in the late eighteenth century and was consistent with the corpuscular nature of light, as was described by Newton in his book *Optika* in 1707 (see Section 4.5). It was recognized through the work of Young and Fresnel in the early nineteenth century that light is a wave with zero mass. Even in the photon picture of light quanta, light is massless. As such light does not feel the force of gravity and should be able to escape no matter how strong the gravitational pull of the object. There is therefore no chance of forming a black hole—light would always be able to escape. This is the picture in Newtonian mechanics.

Black holes, as understood today, are a very different kind of objects. They are point-like objects whose space and time properties are very different from all the other known objects in the universe. The gravitational pull is so strong that even massless light particles, photons, cannot escape such black holes. They were first predicted on the basis of the Einstein's general theory of relativity in 1915. Einstein was initially reluctant to accept this amazing consequence of his theory. The term "black hole" was coined by the American physicist John Wheeler many years later in 1967. The first black hole was spotted in 1971. Since then, a large number of stellar objects have been identified as black holes. The number of black holes in the universe runs into trillions. In our own galaxy, the Milky Way, there are about one billion black holes. The closest known black hole to earth is named Cygnus X-1 and is about 6000 light years away from our planetary system. There is a massive black hole formed at the center of almost all galaxies, including our own.

17.1 General theory of relativity and black holes

When Albert Einstein presented the general theory of relativity in November 1915, it was formulated in terms of a very complicated set of mathematical equations. Given a distribution of mass and energy, it could predict the space-time curvature everywhere in the universe. In general, it is very difficult to find the mathematical solutions of Einstein's equations of gravity. People are still discovering the solutions of these equations for different situations.

The first person to solve Einstein's equations was Karl Schwarzschild. Within weeks of Einstein's formulation of general relativity equations, Schwarzschild solved the simplest problem of a spherical object with a certain mass and radius. He assumed that the object was static and not rotating. Based on the solution, now known as the Schwarzschild metric, he could calculate the space-time curvature in the vicinity of this object and predict all the effects like gravitational time dilation (time moves slower in regions where the gravitational pull is larger) and gravitational red shift (light's frequency decreases toward the red end of the spectrum as we move away from the massive object). One of the most startling predictions based on Schwarzschild's calculation was the existence of black holes.

According to Schwarzschild's solution, if a massive object, like a star, collapses under its own weight then, under certain conditions, it can reach a critical radius. Beyond this radius, no force, no matter how large, can stop further collapse. The star eventually collapses to a point-like object, called a *singularity*. A *singularity* is smaller than the smallest object we can imagine. A black hole is formed.

The picture of a black hole is thus of a point-like singularity with a very large mass concentrated in it. This singularity is surrounded by a sphere with nothing in it, a hole, such that anything, even massless photons, falling into this sphere are not able to escape and collapse into the singularity. The radius of this sphere is called the *Schwarzschild radius* and the surface of the virtual sphere is called the *event horizon*. The event horizon is not a real surface, it is an artificial surface that surrounds the collapsed star and represents the boundary of the black hole. The gravitational force is finite just outside this surface but it is infinite at the event horizon. The prediction of this singularity is amazing as the laws of physics that apply to all other objects in the universe fail to apply at this singularity. This is the place where density is infinite, space is infinitely curved, and time stops.

How to understand the formation of a black hole? Any stellar object with enough mass can collapse and become a black hole. A spherical object with a certain mass, like a star, has a gravitational force acting toward the center. If the star starts collapsing (as discussed in Section 14.2) under its own weight, the density can become large and the gravitational force increases. For example, the force of gravity due to a star would increase four times if it collapses to a sphere of half the star's original radius, sixteen times if it collapses to one fourth of the star's radius, and so on. The force of gravity increases as the size of the dying star decreases. According to the theory of relativity, if a star of a certain mass collapses to a certain radius, the Schwarzschild radius, then the gravitational collapse becomes impossible to stop. The force required to stop the collapse becomes infinite. The star keeps getting smaller and eventually collapses to a point, the singularity. The force of gravity becomes infinite and nothing can escape. Not even light!

How do we understand that light cannot escape a black hole? As discussed in Section 16.2, time moves slowly in regions close to a massive object where the gravitational pull is large. This leads to the gravitational red shift effect: the wavelength of the light signal sent from the region of close to a massive object increases as a result of gravitational time dilation. The infinite gravitational pull of the black hole implies that time stops there. Therefore, the light signal is infinitely red shifted meaning that a signal of any wavelength would increase to an infinite wavelength. An infinite wavelength means a flat signal with no color whatsoever. This means that if a light signal of any wavelength is sent from a black hole, it will not be seen by an observer outside the black hole. The light never leaves the black hole. An outsider can never see anything beyond the event horizon.

The situation can be visualized via the curved space-time picture as shown in Figure 17.1. As discussed in Section 16.2, the space-time surrounding a massive object is curved. If the mass density increases, the bending becomes larger. For example, the bending is steep in the vicinity of a neutron star. When a black hole is formed the entire mass is concentrated in an infinitely deep and point-like region, a singularity, where the density is infinite.

What happens when more matter falls into the black hole? The matter is attracted toward the singularity and the black hole becomes more massive. There is no limit. In spite of an infinite pull at the event horizon of a black hole, black holes do not suck everything up like a vacuum

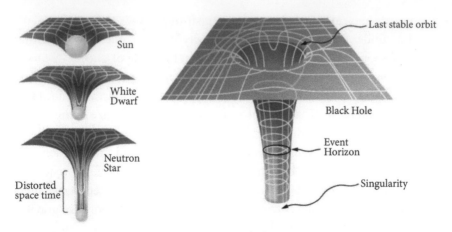

Fig. 17.1 A demonstration of what objects of various masses do to the fabric of space-time. The gravity of these objects bends space-time in three dimensions, but it is not easy to illustrate. For more discussion see Section 16.4.

(Credit: NASA)

cleaner. For example, if the sun was a black hole, all planets would have kept revolving around the sun like before, except that earth would have become cold and dark.

An interesting question is what happens if someone jumps into the black hole? Near the black hole the gravitational pull is extremely strong. A person in an upright position near the event horizon would experience a very different, but extremely strong gravitational pull on different parts of the body. The gravitational pull at the feet will be much larger than the gravitational pull at the head. As a result, the person would be stretched into a long, thin strand of particles before even reaching the event horizon, a terrible fate called "spaghettification."

The nature of this singularity needs a description in terms of quantum mechanical laws as quantum mechanical laws may not be compatible with a point-like object as seen in our discussions of uncertainty laws in Chapter 9. This means that a black hole should be described via a unified theory of general relativity and quantum mechanics. However, despite tremendous efforts, this unification has not been achieved so far. We will not know the true nature of black holes until quantum mechanics and general relativity have been fully integrated as a unified theory.

17.2 How is a black hole formed?

As discussed in Section 14.2, a black hole is the last stage in the life of a massive star—a star whose mass is greater than five times the mass of the sun. When the star cannot produce energy in its core as it has exhausted most of its hydrogen in fusing it into helium, gravity can overcome the balancing force due to the heat produced by nuclear fusion. This causes the core to collapse. The outer layers of the star can, however, be blown away, forming a supernova. The collapsing core produces more heat causing helium to fuse into heavier elements, up until iron. At this stage, according to the prediction of Einstein's theory of gravity, the core collapses and continues to do so if the mass is larger than five times the mass of the sun. Ultimately, when the radius becomes equal to the Schwarzschild radius, gravity becomes so strong that nothing, absolutely nothing, can stop the collapse. Eventually the whole massive core collapses to a point and the black hole is formed. This is the final stage of a massive star. Less massive stars end up being neutron stars as discussed in Section 14.2.

If there is nothing in the neighborhood of a black hole, it will just sit there. However, if there is any gas or dust generated during the supernova stage, the material will be drawn in by the black hole. During this process, there can be bright bursts of light due to the heating of the gas and dust. This is similar to when water swirls around while going down the drain. The black hole becomes more massive in the process.

Supermassive black holes are known to produce quasars which are intensely bright sources of radio wave emissions. We recall from the discussion in Section 4.7 that a radio wave is just like light but with a much larger wavelength that is not visible to the naked eye. Quasars are produced when massive objects fall into a black hole. During the fall, these objects are strongly compressed and get heated up. The result is that there is a glow surrounding the black hole in the form of an accretion disk, a disk formed by diffuse material in an orbital motion. This is shown in Fig. 17.2. In some cases, this glow can be brighter than the entire host galaxy.

A question of interest is: "What is the size of a black hole?" As far as the mass is concerned, all of it is concentrated in a point of

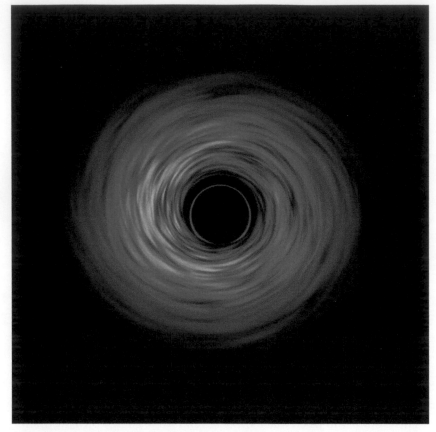

Fig. 17.2 Accretion disk around a black hole.
(Credit: NASA)

negligible volume. However, the size of a black hole corresponds to the Schwarzschild radius or the distance from the singularity to the event horizon. Any object in this region will be swallowed by the singularity. The size of the black hole depends on how massive it is. A black hole formed out of a star five times the size of the sun would have a radius of about 16 kilometers. Typically, the black holes found in our galaxy have 10 to 100 times the mass of the sun. However, there can be supermassive black holes as well. For example, the black hole at the center of the Milky Way galaxy is about 4.3 million solar masses.

Next, we address the question about how to see a black hole. In principle, nothing, not even light, can escape the black hole. This makes it impossible to see a black hole directly. However, observing the gravitational effects of black holes on other objects makes it possible to indirectly infer their presence.

Indirect evidence of black holes comes by witnessing stars orbiting an invisible object. For example, the massive black hole at the center of Milky Way galaxy was inferred by the movements of the neighboring stars.

Another method of an indirect observation of a black hole is via a quasar. Quasar is short for "quasi-stellar radio source" because no one knew the origin of these bright objects when they were first discovered in the 1950s. As discussed above, a quasar is, in some sense, the visible face of a black hole. All black holes do not produce quasars, only supermassive black holes that are millions or billions of times of the mass of the sun can do. A quasar usually has a short life. They outshine the host galaxy until the infalling matter has all been "eaten" by the black hole. More than a million quasars have been found. The nearest known quasar was found to be about 600 million light years away. However, quasars as old as 13 billion light years away have also been found.

The detection of gravitational waves, as discussed in Section 16.5, provides another tool to detect a black hole. When two black holes come close to each other, the intense gravitational force pulls them toward each other. This shakes the fabric of space-time surrounding them, producing gravitational waves. These waves are extremely weak but measurable. A number of black hole merger events have been recorded so far.

17.3 Evaporation of black holes

As mentioned above, one of the amazing predictions of Einstein's general theory of relativity was the existence of black holes—objects so massive with a gravity pull so large that everything falls into them and nothing, not even light, can escape. Everyone believed that black holes, unlike any other object in the universe, will continue to exist for ever. For a long time, this picture of black holes continued to exist.

Then, in 1974, Steven Hawking came up with the astounding conclusion that this may not be the correct picture—black holes will eventually evaporate by emitting massless particles and cease to exist, albeit after an unimaginably long time. Black holes are not entirely black. This is a startling discovery. Black holes that were understood to devour everything could lose mass, dramatically changing cosmological evolution and the final state of the universe.

This result is one of the rare instances where two of the major theories of physics, quantum mechanics and the general theory of relativity, come together to predict something amazing.

In spite of their immense successes, the general theory of relativity and quantum mechanics have maintained their independence and no effort to join them into a single theory has succeeded so far. This remains a long-cherished dream. The difficulties in achieving this goal appear to be insurmountable. In view of this background, it is very impressive that Hawking could come up with this unexpected result.

Steven Hawking was born in 1942. He is regarded as an iconic figure of our time. An aura of brilliance and excellence surrounded him. His medical situation where he was confined to a wheelchair, unable to write or speak, coupled with his brilliant scientific achievements made him a household name, almost as popular as Isaac Newton and Albert Einstein.

The detailed calculations leading to Hawking radiation are too complicated to present here. However, a simple picture, though not completely accurate, helps in understanding the potential death of a black hole.

The esoteric nature of a vacuum is discussed in Chapter 9. A vacuum is a sea of an infinite amount of energy. Pairs of particles and antiparticles keep popping into and out of the vacuum. In the region just outside the event horizon, such pairs are generated. Typically these pairs recombine immediately and are converted back to energy. However, once in a while, one of the particles of the pair enters inside the event horizon before it can be annihilated by its partner. The one inside the event horizon can no longer escape due to the infinite pull of the black hole. However, the partner left outside can escape carrying a certain amount of energy. The question now is where does this energy come from? The source of energy surrounding the black hole is the black hole itself, just like an object near earth has gravitational energy due to the mass of the earth. Thus, it follows from the law of conservation of energy that the lost energy comes from the black hole and the mass of the black hole decreases according to the Einstein relation $E = mc^2$ that equates energy and mass.

This simple argument rooted in the quantum fluctuations of the vacuum and the properties of black holes based on the general theory of relativity and combined with the energy–mass equivalence explains how black holes will completely evaporate someday. The timescale over which the evaporation of a black hole takes place is several trillion times the age of the universe and none of us should expect to see this amazing consequence of the quantum mechanical description of gravity in our lifetime.

17.4 Rotating black holes and the possibility of time travel

In 1915, Albert Einstein put forward the equation that governs the structure of space-time in the presence of massive objects. As mentioned earlier, within weeks, Karl Schwarzschild found the first exact solution of the Einstein equation for the simplest object: a massive sphere with no rotation and no electric charge. This solution led to the prediction of the black hole, a point-size massive object surrounded by an event horizon. Almost fifty years later, in 1963, Roy Kerr found another exact solution of Einstein's equations, this time for a spherical object that is rotating. The mathematical details are very complicated and the results are highly unusual. A prediction of the Kerr solution is the existence of a novel and exotic black hole.

Rotating black holes are among the greatest wonders of this universe. They are formed when a rotating star collapses beyond an event horizon. They have dramatically different properties than the Schwarzschild black hole as depicted in Figure 17.3. The first major difference is that, instead of a point singularity, the rotating star collapses into a ring forming a ring-shape singularity. Any falling mass in the rotating black hole will be concentrated on the ring. The size of the ring depends on the mass and the rotation rate. The second big difference is that a rotating black hole has not one but two event horizons. The inner event horizon is the

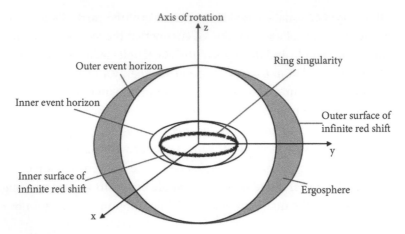

Fig. 17.3 Schematics of a rotating black hole. Instead of a point, the mass is concentrated on a ring leading to a ring singularity. In addition, there are two event horizons that form the outer surfaces of ergospheres.

spherical surface surrounding the ring from which nothing can escape. A second event horizon forms the outer surface.

In addition to the event horizons, there is an outermost surface called the stationary limit. Beyond the stationary limit, an object can remain in a fixed position. The region between the outer event horizon and the stationary limit is called an ergosphere. The ergosphere is an ellipsoid region around the black hole. A particle falling into the ergosphere can escape as long as it does not hit the inner event horizon. If the object crosses that event horizon, it will be drawn into the black hole and never escape.

According to the general theory of relativity, the space around a rotating massive object is not static or fixed with respect to the rest of universe. That space is dragged around the object. The rotating black hole can drag that space around itself at speeds faster than the speed of light in the ergosphere. An object falling into the ergosphere experiences enormous accelerations, gaining energy from the black hole's rotation. It can still escape the gravitational pull of the black hole but is dragged along with the swirling space-time. The swirling effect is so intense that even light cannot advance against the rotation in this region. The powerful dragging of space-time due to the rotating black hole can provide a mechanism for jets. Objects caught inside the ergosphere can acquire tremendous energy with relativistic speeds before being ejected as a jet in a spectacular fashion.

In Fig. 17.4, an artist's concept of a supermassive rotating black hole is illustrated. The black hole is surrounded by an accretion disk consisting of the debris falling into the black hole as discussed above. In addition, there is an outflowing jet of fast-moving particles that is caused by the spin of the black hole.

Einstein's theory of relativity predicts the existence of wormholes as a consequence of the curving of space-time surrounding a massive object. A wormhole is a tunnel through space-time that connects different parts of the cosmos.

In order to illustrate this concept, let us consider an example of a two-dimensional space as it is easy to visualize. Let the two-dimensional space be in the form of a sheet which is folded as a result of some distribution of massive objects as shown in Fig. 17.5. Now a massive object at a point on the upper part can cause a space-time curvature. If a mass is placed on the lower side but opposite to the object on the upper side,

Fig. 17.4 This artist's concept illustrates a supermassive rotating black hole. A black hole can hold millions to billions of times the mass of the Sun. (Credit: NASA)

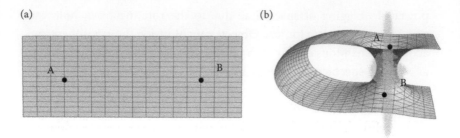

Fig. 17.5 A wormhole is a tunnel through space-time that connects different parts of the cosmos. In (a), the two points A and B can be hundreds of light years away in a flat space. In (b), the same two points A and B can be very close in a curved space forming a wormhole.

space-time is curved in the opposite direction. A space-time tunnel is produced. This is called a wormhole.

In the universe, there is a possibility that different masses can create the space-time structure corresponding to a wormhole. This leads to the possibility of entering a wormhole and coming out at a different time, past or future, thus making time travel possible.

18

Big bang—Birth of the universe

It is a remarkable thing to be able to say just what the universe was like at the end of the first second or the first minute or the first year. To a physicist, the exhilarating thing is to be able to work things out numerically, to be able to say that at such and such a time the temperature and density and chemical composition of the universe had such and such values.

—*Steven Weinberg*

How and when was our universe created? This is the question that has perplexed humans since antiquity. Many scholars, philosophers, and scientists have tried to answer this question. This question is also central to many religions. Humans have appreciated the vastness of the universe and tried to understand its origins, limits, and nature. The discoveries of the last century have brought us closer to an understanding of the beginning and the evolution of the universe. However, whereas we feel more educated about the origins of the universe, we have many fundamental questions unanswered. The mysteries about the nature of space and time abound.

The presently accepted model is the so-called big bang model. According to the model, the universe came into being as a result of a big bang about 13.8 billion years ago. It seems unbelievable but, at that moment, all the trillions of galaxies and the clusters of galaxies with each galaxy consisting of billions of stars were concentrated at a single point, a point much smaller than the smallest object we can think of. Space and time were created as a result of an expansion of that point and the universe, as we see around us, came into being. The universe has been expanding ever since. While the spatial size of the entire universe is unknown, there are indications that, at present, it has a diameter of several trillion light years. The size of the observable universe is estimated to be approximately equal to 93 billion light years in diameter. Whether the universe is finite or infinite remains an open question.

The big bang model of the universe is supported by a number of important observations. The first was the observation by Edwin Hubble in 1929 that far-away galaxies are receding from us. The expansion of the universe provided the first clue that the universe may have a beginning. Another important piece of supporting evidence came in 1964 when Arno Penzias and Robert Wilson discovered that the universe is immersed in a background radiation that is thought to be left over from the first moments of this universe. Finally, the observation that light atoms like hydrogen and helium constitute most of the observed universe provided support for the big bang theory. These atoms were formed when the protons and neutrons that were generated during the first few minutes of the birth of universe fused together. These three measurable signatures strongly support the notion that the universe evolved from a dense, nearly featureless hot gas, located at a single point.

In this chapter, we explore how this exotic and incomprehensible picture came to be accepted. What is the evidence to support such an explosive beginning? How does this model explain the present status of the universe? And what can we predict about the ultimate fate of the universe? These are truly perplexing questions.

18.1 Expansion of the universe

As discussed in Chapter 16, when Einstein formulated the general theory of relativity in 1915, our cherished concepts of space and time were overturned. One of the most startling consequences of the theory of relativity was that the universe was not static and was expanding. This consequence was so radical to the existing belief that the universe was static and infinite that Einstein himself could not reconcile with it. His unease with the concept of an expanding universe is demonstrated by the fact that he artificially included a term, called the cosmological constant, in his equations to ensure a steady universe. This was something that he would regret later and regard it as the biggest blunder of his life.

Why was the idea of an expanding universe so unacceptable? An expanding universe implied that, projecting back into the past, the universe must have been like a point a long time ago.

The first person to predict the expansion of the universe was a Russian scientist, Alexander Friedmann. In 1922, he derived a set of equations from the general theory of relativity, after neglecting Einstein's cosmological constant, which showed that the universe might be expanding. Friedmann did not live long and died in 1925 before his prediction was shown to be true. In 1927, Georges Lemaitre independently derived similar results.

One of the greatest scientific discoveries of the twentieth century is the observation by Edwin Hubble that the universe, filled with trillions of galaxies, is expanding at a fast rate. This simple observation would change our perception of the universe forever. This work led to the birth of the new field of observational cosmology with amazing consequences. The most dramatic was the realization that the universe had a beginning.

In a research paper published in 1929, Hubble showed that galaxies far away are receding from us, the receding speed being directly proportional to the distance from us. Thus, a closer galaxy is receding with a smaller speed as compared to a galaxy farther away. Hubble came up with this astonishing conclusion on the basis of observational data of 24 galaxies for which the distance as well as the speed with which they are moving away from earth were available.

The idea of an expanding universe can be understood by a simple example. Let us visualize our universe to be like a dough with galaxies embedded in it like raisins inside the dough as shown in Figure 18.1. The raisins are distributed randomly inside the dough with fixed distances from each other. As the dough rises or expands (as during baking), the raisins move farther away from each other, From the perspective of a raisin (say raisin B), raisin A which is closer to raisin B moves slower as compared to raisin C which is farther from raisin B. As shown in Figure 18.1, if the initial distances of raisin B from the raisins A and C were 5 centimeters and 10 centimeters, then in the expanded dough these distances become 10 centimeters and 20 centimeters, respectively. If this expansion takes place in (say) 10 minutes, then raisin A will be moving away from raisin B at a speed of 5/10 = 0.5 centimeter per minute

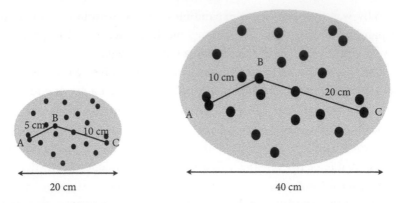

Fig. 18.1 The dough with the raisins. As the dough rises, the distance between the raisins increases. The farther raisins move faster as compared to the nearer raisins.

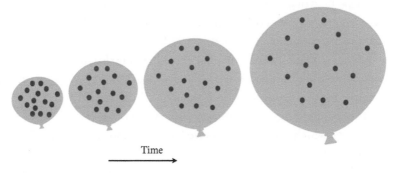

Fig. 18.2 The surface of the balloon is assumed to be two-dimensional universe with dots representing the galaxies. As the balloon expands, the galaxies move apart as a consequence of the expansion of the "space". The farther the galaxy, the faster it moves away.

whereas raisin C will be moving away at the speed 10/10 = 1 centimeter per minute. This explains how the closer galaxies are moving slower and distant galaxies are moving faster in an expanding universe.

As another example, we can visualize a two-dimensional universe as the surface of a balloon with points representing galaxies spread randomly on its surface. As the balloon expands, the distance between the galaxies increases (Figure 18.2). For any galaxy on the surface, the same outcome is obtained—the closer galaxies move away more slowly as compared to the galaxies that are initially farther away.

From the astronomical data, Hubble concluded that the ratio between the relative speed of the galaxies and the distance between them was constant. This constant, perhaps the most important in the field of cosmology, is named in Hubble's honor as the Hubble constant. Hubble calculated it to be 170 kilometers per second per million light years, that is, for galaxies one million light years apart, the difference in their speed is 170 kilometers per second.

In order to understand the Hubble constant, we consider a simple example. Let us consider two cars A and B. A is moving at 30 kilometers per hour and B is moving at 60 kilometers per hour. If the distance between them is 300 kilometers, then the corresponding "Hubble constant" will be the relative velocity (60 – 30) = 30 kilometers per hour divided by the distance between the two cars equal to 300 kilometers. This comes to 30 kilometers/hour per 300 kilometers or 1/10 kilometers per hour.

It turns out that Hubble was off by a big amount in his estimation of the Hubble constant for the universe. Over several decades, a more accurate value for the Hubble constant has been obtained and it is now set at 15 kilometers per second per million light years. Evidence for the expansion of the universe was a shock, particularly for those who believed in a steady-state universe.

Hubble's law had a close resemblance to Kepler's laws. Both of them were empirical laws based solely on observational data. Just as Kepler's laws could be explained by Newton's law of gravitation, Hubble's law relating two measurable properties of the galaxies, their speed and their distance, was a consequence of Einstein's theory of gravitation.

The startling observation is that galaxies in space are not moving away from each other, it is the space between the galaxies that is expanding. How to understand that? Going back to the example of an expanding balloon, the dots representing galaxies on the surface of the balloon recede from each other, not because they move away from each other, but because the space between them is expanding. The apparent speed appears to be larger for the more distant galaxies.

Fig. 18.3 If a galaxy is receding away from earth, the light emitted by the stars inside the galaxy are red-shifted. The red-shift is proportional to the speed with which the galaxy is moving away. This is the way the speed of the galaxies is determined.

A question of interest is: How can we find out that the universe is expanding? The basic idea used in these measurements is the red shift of light coming from distant galaxies. The light emitted by a star in a distant galaxy has a certain wavelength. If all the galaxies were static, then this light would have been received with no change. However, if the galaxy is moving farther away, the space between the galaxy and the earth is increasing (Figure 18.3). The light waves get stretched and the wavelength gets larger. Thus, (say) the blue light emitted by a receding star would tend to turn to red due to the expansion of the universe. The amount of the wavelength shift can tell us how fast the rate of expansion is. The cosmological red shift is a standard method for finding the motion of distant objects.

18.2 How old is the universe?

The most important consequence of the expanding universe is that, if the universe is expanding at a constant rate, there should be a moment in the past when the entire universe was localized at a single point—the universe must have a beginning! ! ! This was a startling realization.

How to find the age of the universe from Hubble's law?

Let us go back to the example of two cars A and B that are moving in the same direction. We ask the question: if A is moving at 30 kilometers per hour and B is moving at 60 kilometers per hour and they are 300 kilometers apart, how long ago were they at the same point? The answer is 10 hours. One way of seeing this is that, if both cars started at the same time at some location, then, after 10 hours, car A moving at 30 kilometers per hour will have moved 300 kilometers, whereas, in the same duration, car B moving at 60 kilometers per hour will have moved 600 kilometers. The separation between the two cars will be 300 kilometers.

Another straightforward way to answer this question about how long ago the two cars were together is to take the inverse of the Hubble constant. In our example, the Hubble constant is 30 kilometers/hour per 300 kilometers and the answer is 300/30 = 10 hours.

Now we are ready to answer the question: How old is the universe? If galaxies that are one million light years apart are moving with a relative speed of 15 kilometers per second, the time when they were together was one million light years ago divided by 15 kilometers per second. This comes to 20 billion years.

This is somewhat different from what we believe to be the actual age of the universe which is estimated to be 13.8 billion years. The reason for this discrepancy is that the age calculated from the Hubble constant assumes that the expansion rate of the universe has been the same from the moment when the big bang took place till the present time. However, this is not true. For example, as we discuss later in this chapter, the universe expanded immensely fast in the initial moments and then the expansion rate slowed down. The expansion rate at any time also depends on the energy-mass content of the universe at that time. For example, the galaxies may have been slowing down under the influence of their mutual gravitation.

Next, we ask the question: Is there other evidence for the big bang? There is more. Perhaps the most convincing is the discovery, in 1964, that there is radiation left from the moment that this universe was born, the cosmic microwave background radiation.

18.3 Cosmic microwave background radiation

The discovery of cosmic microwave background radiation has a colorful history. In 1964, radio astronomers Arno Penzias and Robert Wilson were measuring the intensity of radio waves emitted by stellar objects in our galaxy, the Milky Way. The Milky Way, as we have seen, is a planar galaxy with a small hump in the middle. Their particular interest was looking in the regions out of the plane of the galaxy. These observations were being made using a radio antenna at Bell Telephone Laboratory. A radio antenna is a very sensitive device.

What Penzias and Wilson found was that they received microwave radiation signals no matter which direction they pointed their radio antenna. Initially they thought there was something wrong with their antenna. They noticed that a pair of pigeons were roosting within the antenna. The antenna was coated, in the delicate words of Penzias, by a "white dielectric material." This could potentially be the source of the unwanted microwave radiation. The antenna was dutifully cleaned. But this did not change things, the microwave radiation was still there. A careful measurement revealed that the equivalent temperature was about 3.5 degrees Kelvin. The mystery faced by Penzias and Wilson was: What is the source of this radiation?

What Penzias and Wilson did not know was that few miles away, at Princeton University, a young scientist, Phillip Peebles, was predicting that, in the first few minutes of the big bang, there must have been an enormous amount of radiation that must have been cooled to a few degrees Kelvin by now. Peebles' ideas were inspired by a senior colleague at Princeton, Robert Dicke, who also thought that there was some radiation left from those earliest moments of the universe. Dicke was planning to build a radio antenna to search for this left-over radiation. This is when he and Peebles became aware of Penzias and Wilson's observations and explained to them that what they were observing was this background radiation.

Penzias and Wilson were awarded the 1978 Nobel Prize. Peebles had to wait a long time until he was awarded the 2019 Nobel Prize.

The origin of these signals can be understood as follows.

The universe started out from a point with an enormous amount of radiation. The expanding universe was dominated by charged particles, electrons and protons, and electrically neutral neutrons as well as light particles, photons, during the first 380,000 years of its life. This was the time when the universe was extremely hot. It was also opaque as light was strongly scattered from the electrons and protons and did not travel much farther. The soup of these charged particles acted like cosmic fog.

The situation was similar to light scattering inside a thick fog—light bounces from particle to particle and does not leave the fog. When the temperature of the universe cooled as a consequence of its expansion to about 3,000 degrees Kelvin, these particles, electrons, protons, and neutrons, started forming atoms. Since light does not interact strongly with atoms, it was able to travel long distances—the universe became transparent (Fig. 18.4). This situation persists to this day and we are able to see light travelling billions and trillions of light years without hinderance.

We discussed in Chapter 3 how Max Planck had discovered the relationship between wavelength and temperature. He showed that, if the energy of a radiation field is constant, then the temperature is inversely proportional to the wavelength of light (distance between the crests of the light wave). This means that, when the temperature increases, the wavelength decreases and vice versa. The temperature of 3,000 degrees Kelvin corresponds to the light distributed around a wavelength of one micron (a millionth of a meter). The effect of the expansion is to increase the photon wavelength in proportion to the size of the universe.

This can be seen from Figure 18.5. Here we draw a wave on an elastic band. If the elastic band is stretched, the wave gets stretched and

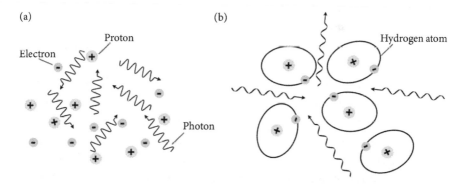

Fig. 18.4 (a) During the first 380,000 years since the big bang, the universe consisted mainly of charged particles like electrons and protons. Light was trapped in this soup of charged particles. (b) When, at the end of this period, the universe cooled, electrons and protons joined together to form hydrogen and helium atoms. Light could now travel long distances and the universe became transparent.

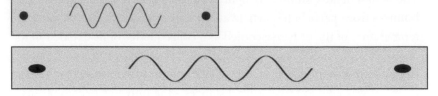

Fig. 18.5 As the elastic band is stretched, the wave gets stretched and the wavelength increases.

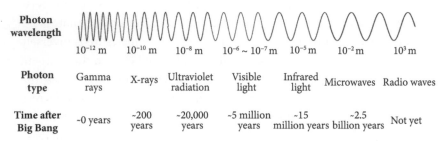

Photon wavelength							
10^{-12} m	10^{-10} m	10^{-8} m	$10^{-6} \sim 10^{-7}$ m	10^{-5} m	10^{-2} m		10^{3} m
Photon type	Gamma rays	X-rays	Ultraviolet radiation	Visible light	Infrared light	Microwaves	Radio waves
Time after Big Bang	~0 years	~200 years	~20,000 years	~5 million years	~15 million years	~2.5 billion years	Not yet

Fig. 18.6 As the universe expands, and consequently cools, the wavelength of the cosmic background radiation increases. At present this background radiation is in the microwave region.

the wavelength increases. Similarly, when the universe expands, space expands and the wavelength of the light or the photons increases. The rate of increase in the wavelength is directly proportional to the rate of expansion of the universe.

The universe has expanded by 1,000 times since the time the temperature of the universe was 3,000 degrees. Therefore, the photon wavelength has increased by 1,000 times. The present wavelength of the left-over radiation is therefore equal to one micron multiplied by 1,000 (a thousandth of a meter or 1 millimeter). This corresponds to the microwave radiation. The corresponding temperature is equal to 2.87 degrees Kelvin. This is the cosmic background microwave radiation that Penzias and Wilson detected in 1964. This is the most ancient signal received by astronomers. The change in the wavelength of the cosmic background radiation as it expands and cools is given in Fig. 18.6.

Temperature distribution is uniform throughout the universe. It was predicted that there should be small temperature variations in the microwave background radiation as the afterglow of the big bang

Fig. 18.7 The temperature distribution in the universe.
(Credit: ESA/Planck)

(see Fig. 18.7). These variations are so small, one part in ten thousand, that they could not be detected from earth. In 1992, NASA sent a cosmic background explorer (COBE) satellite to scan the entire sky to search for the temperature variations. The resulting measurements were in accordance with the theoretical predictions and provided solid evidence for the validity of the big bang model. These tiny temperature variations correspond to the largest-scale structures of the observable universe. A region that was a fraction of a degree warmer becomes a vast galaxy cluster, hundreds of millions of light years across.

18.4 Big bang: The first second

The universe started with all its mass and energy concentrated at a single point. At that point there was no space and no time. It is hard to imagine that the entire universe we see around us was concentrated in a single point. And then the big bang happened. Unlike the name, the big bang did not start with an explosion with a large sound—it started with a rapid expansion. What prompted this beginning is still a mystery.

The first second of the universe is the most important in the life of the universe. What happened in these earliest moments essentially determined what kind of universe it would be and what would be the laws of nature that will govern all the matter and energy contained in this universe. Here we present the evolution of the universe in that critical second based on the existing theories.

The earliest moments of the universe extend from the moment the big bang took place to 10^{-43} seconds.[1] This is an incredibly short time, equal to a billionth of a trillionth of a trillionth of a trillionth of a second. This period was dominated by the quantum fluctuations which are discussed in Chapter 9. This era is therefore appropriately referred to as the Planck era, named after the founder of the modern quantum era. What happened in the Planck era cannot be fully understood until the two major theories, quantum mechanics and the theory of relativity, are combined into one unified theory.

By the end of the Planck era, the universe had expanded to 10^{-35} meters in diameter. This is an extremely small distance, trillions of times smaller than the size of an electron. All the forces of nature, the gravitational force, the electric as well as magnetic forces, the nuclear force that binds the nucleus together, and the weak force that allows emission of particles from the nuclei in the process called radioactivity, were all unified in this epoch. This is shown in Fig. 18.8.

At the end of the Planck era, the gravitational force became detached from the other forces. During this period, the laws that would eventually govern the universe were established. During a time from 10^{-35} seconds to 10^{-32} seconds, the universe experienced an incredible burst of expansion. Most cosmological models suggest that the universe at that point was filled homogeneously with a high-energy density and the rapid expansion was caused by an incredibly high temperature and pressure. In this period, called the inflation era, the volume of the universe increased by a factor of 10^{78}. The inflation era is discussed in some detail in a later section.

During this time, the universe was dominated by energy in the form of photons and subatomic particles like electrons and quarks. Quarks

---

[1] Recall that 10^{-43} is a compact notation for 1/10,000,000,000,000,000,000,000,000,000,000, 000,000,000,000

Similarly 10^{78} = 1,000,000,000,000,000,000,000,000,000,000,000,000,000,000,000,000,000,000, 000,000,000,000,000,000,000,000

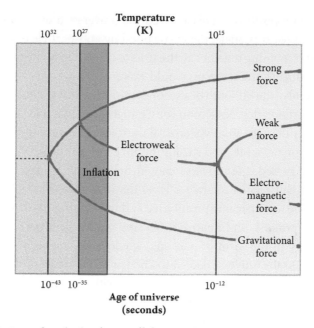

Fig. 18.8 Just after the big bang, all forces of nature were unified. Over the period of less than one second, these forces became independent of each other.

are exotic objects that have never been observed directly. They would eventually combine together to form the protons and neutrons that constitute the nuclei of all the atoms in the observed universe. However, in this early period, the universe was extremely hot with enormous energy. This energy was producing massive particles and their anti-particles, such as pairs of electrons and positrons. A positron is the anti-particle of an electron. This conversion of energy and mass was according to the law of energy-mass equivalence as discovered by Einstein as a consequence of the theory of relativity (Section 15.6). Most of these particles and their anti-particles would then recombine to convert back to energy. This exchange of energy and massive particles continued during this era of unbelievably high temperature.

This is the period when something happened that resulted in the anti-particles like positrons being completely dominated by particles like electrons, thus leading to a high asymmetry in the number of (say) electrons and positrons. This allowed the universe to exist as we see it today. If, for each particle, the corresponding anti-particle existed, they would be combining together all the time to convert to photons. No stars, planets, humans, trees, oceans would exist.

At the end of the billionth of a second, the universe had cooled to about thousand trillion degrees Kelvin (or Centigrade) and it had expanded to a size larger than the solar system. All the forces of nature (gravitational, electromagnetic, weak, and strong) became independent.

As time passed, the universe cooled further. By the end of the first second, it had cooled to such a level that the generation of particle–anti-particle pairs became prohibitive and the existing particles, such as electrons, protons, and neutrons, became stable. Such particles, along with energy in the form of photons, would form the seed for further evolution. As the universe cooled to about a billion degrees, it expanded to almost ten light years across. Obviously with an expansion of several light years in just one second, that means that the expansion rate was faster than the speed of light.

The expansion rate had considerably slowed down after the initial inflation era. For example, the universe was the size of our galaxy, the Milky Way, with a radius of 100,000 thousand light years after about three years of its existence.

18.5 Stars, galaxies, and planets are formed

When the universe was one second old, it was too hot to form stable nuclei. At that time, electrons, protons, and neutrons formed a sea of charged particles. The further evolution is depicted in Fig. 18.9.

After a few minutes, the temperature dropped to 1 billion degrees. At this temperature, the protons and the neutrons started to combine and form deuterium nuclei. Deuterium is a form of hydrogen with its nucleus consisting of one proton and one neutron. The deuterium nuclei combined to form helium nuclei, consisting of two protons and two neutrons, via a process called nucleosynthesis. This is shown in Fig. 18.10. After the first three minutes, no new elements were formed for millions of years. The universe consisted only of hydrogen and helium nuclei and energy.

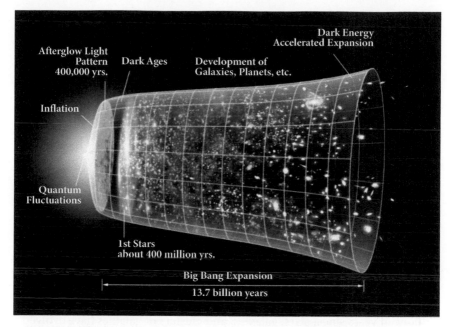

Fig. 18.9 The growth of the universe since the big bang. The big bang was immediately followed by an era of "inflation". The expansion has been slow since the inflationary period.

(Credit: NASA/WMAP)

The universe was still too hot for the electrons to settle around protons and neutrons to form neutral atoms. Until the universe was about 380,000 years old, electrons, positively charged nuclei of lighter elements, and photons dominated the universe. At the end of this period, the temperature was 3,000 degrees. This is roughly the moment when electrons combined with hydrogen nuclei and formed neutral hydrogen atoms. Light (or photons) could now travel long distances—the universe was no longer opaque. This heralded the beginning of the atomic period. Particle generation at different times in the evolution of the universe is shown in Fig. 18.10.

It took a very long time for the force of gravity to bring together clouds of hydrogen and helium to form a star. The first stars were formed about 180 million years after the big bang. The nuclear fusion process in the cores of the stars started. Hydrogen fused into helium, helium fused

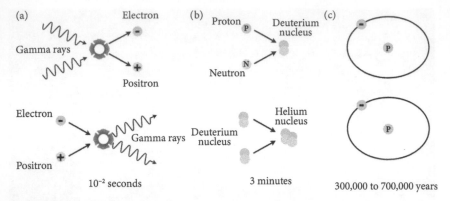

Fig. 18.10 Particle generation at various moments in the life of the universe.

into carbon, and various other elements were formed up to iron. It took another 200 million years before the first galaxies were born. In the billions of years since, stars, galaxies, and clusters of galaxies have evolved into other cosmic objects like pulsars, neutron stars, and black holes.

The galaxy formation process still continues. The galaxies can merge together under the gravitational force of attraction. For example, our galaxy, the Milky Way, is about 2.5 million light years away from the nearest galaxy, the Andromeda galaxy. In spite of such a large distance, there is a gravitational pull between the two galaxies and they are approaching each other at a speed of nearly a quarter million miles per hour. At this rate, they are expected to collide in about 3.75 million years (see Fig. 18.11). This appears quite surprising. How can the gravitational pull be so strong as to attract a galaxy so far away? It turns out that the galaxies are extremely massive—the Milky Way contains 300 billion stars and the Andromeda galaxy has about one trillion stars. Although the individual stars within a galaxy can also be far away from each other (the nearest star to the sun is Proxima Centauri which is 4.5 light years away), the galaxies are truly massive with a sufficiently large overall density. If the galaxies are considered a homogeneous distribution of mass, it is understandable that they exert a strong force of gravitation upon each other. Indeed, the galaxies, though quite far apart, can be quite crowded on the galactic scale. For example, the Milky Way is

Fig. 18.11 An artist's depiction of the collision of the Milky Way galaxy (right) and its neighbor the Andromeda galaxy (left). This is how the sky may look 3.75 billion years from now. The Milky Way may get distorted due to the tidal pull of the Andromeda galaxy.

(Credit: NASA; ESA; Z. Levay and R. van der Marel, STScI; T. Hallas; and A. Mellinger)

about 100,000 light years across. As such, the distance to the Andromeda galaxy is only about 25 times the size of the Milky Way galaxy.

Our sun was born about 9 billion years after the big bang. The solar system, including earth, came into being at that time. Since the universe is now thought to be 13.8 billion years old, that makes our sun about 4.8 billion years old.

The earth was born 4.6 billion years ago. For the first 700 million years, earth was simply too hot to sustain any kind of life. The temperature gradually cooled and the earth became more hospitable around 4 billion years ago. The earliest rocks with fossil evidence of life found on earth are about 3.5 billion years old, more than 10 billion years after the creation of the universe. In some sense this was an epoch-making event. Just imagine that, for all those billions of years, there was no earthly creature to observe the universe or make use of it.

A great mystery is how life started on our planet, earth. How did all the basic building blocks of life, such as DNA, RNA, amino acids, and sugars, come together to create life? It is still a scientific mystery how life emerged out of non-life. There are many theories about the beginning of life on earth but there is no consensus. One plausible theory is that of a "primordial soup," the idea that the young earth had oceans full of chemicals that were important for life. They would, at some point, self-assemble into the simplest form of life, a living cell. However, the exact route to the beginning of life has eluded scientists so far.

18.6 Shape of the universe

An important question is: What is the shape of the universe?

We have repeatedly argued that a consequence of Einstein's general theory of relativity is that space and time can be warped and have curvatures. The curvature depends on the distribution of massive objects in the universe. It depends on the nature of the curvature whether the universe, as we see it, is closed, flat, or open. These concepts can be made clearer by describing simple two-dimensional analogues as it becomes quite difficult to visualize curved space-time in four dimensions. We see three standard two-dimensional shapes in Fig. 18.12. A spherical surface is an example of a closed surface. The two-dimensional spherical surface curves around such that the surface area remains finite. A sheet of paper is an example of a flat surface. Such a surface can extend to infinitely large areas. And finally, a saddle shape is an example of an open surface. It can also extend out to infinity.

The question about the shape of the universe, whether it is closed, flat, or open, is an important one, for an answer to this question can shed light on the fate of the universe. A simple picture of the universe is as follows. There are two forces acting. Due to the initial big bang whose origins we do not understand, the universe is expanding and the mass density is decreasing. Density is a measure of how much mass is spread over its volume. And then there is the force of gravity. A massive universe has the tendency to collapse under the gravitational force. A balance between these two forces determines whether the universe is closed, flat, or open.

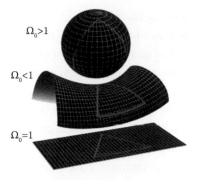

$\Omega_0 > 1$

$\Omega_0 < 1$

$\Omega_0 = 1$

Fig. 18.12 The shape of the universe depends on its density. If the density is more than the critical density, the universe is closed and curves like a sphere; if less, it will curve like a saddle. But if the actual density of the universe is equal to the critical density, as scientists think it is, then it will extend forever like a flat piece of paper. The parameter Ω_0 is a measure of the flatness of the two-dimensional surface: The surface is closed when Ω_0 is greater than 1, it is open if Ω_0 is less than 1, and flat if Ω_0 is equal to 1. (Credit: NASA)

If the density is greater than a critical density, then the universe contains enough mass to eventually stop the expansion and start compressing. The end result is that the expansion due to the initial big bang will come to a stop and a collapse would start under the force of gravity. Eventually the universe would be compressed to the same kind of a point-like singularity it started with. This scenario is called the "big crunch."

If, on the other hand, the total matter in the universe is just about right and the density is equal to the critical density, then the two forces would balance each other and the universe becomes flat—neither expanding nor compressing.

Lastly, if the matter density is below the critical density, then the force of gravity is not sufficient to stop the expansion and the universe keeps becoming bigger and bigger with no end. This is the scenario for an open universe.

Next question is: How to find whether the universe is closed, flat, or open? Again, we consider the two-dimensional examples discussed above. The basic difference between the three can be illustrated with geometrical axioms going back to Euclid. They can, in turn, be useful in experimentally determining the nature of the universe.

The simplest axiom of geometry is that two parallel lines never meet and remain parallel forever. This is however true only for "flat geometry." For curved geometry, this is not true. Next we consider a closed geometry like a sphere. On the surface of a sphere, a line is defined as the great circle that encompasses the sphere. A great circle is a circle whose center lies at the center of the sphere. No matter how they are drawn, each pair of great circles will always intersect. As a result, parallel lines do not exist on the surface of the sphere. As an example, if two lines are drawn at the equator pointing north, they will meet at the north pole. On the other hand, parallel lines drawn on an open surface like a saddle never meet but they diverge. The distance between the two lines increases.

Another elementary geometrical property relates to triangles. A well-known property of triangles is that the sum of the three angles inside the triangle add up to 180 degrees. This statement is ingrained in us since our first lessons of geometry. However, this statement only applies to a flat geometry as well. On a closed surface like a sphere, the three angles add up to more than 180 degrees and on an open surface, they add up to less than 180 degrees. This can be seen in Fig. 18.12. It should however be noted that, in both examples, the results of the "flat geometry" may appear valid in small regions of large objects even when the objects are not flat. For example, the earth appears flat around us although it is spherical.

How do we use these ideas to measure the curvature of the universe?

There are many different ways to determine whether the universe is closed, flat, or open. Perhaps the most impressive clue to the shape of the universe has come from the delicate measurement of the temperature variations of the cosmic microwave background radiation over a very wide region. As discussed in Section 18.3, there are hot and cold spots in the cosmic background radiation about the size of the diameter of the moon spread out throughout the universe. The size of these spots and their distances are known very precisely. It then becomes possible to form a cosmic isosceles triangle with the cosmic background as the base with earth at the apex. If the universe is closed, the "hot" spots appear larger than their actual size. However, these spots appear smaller than

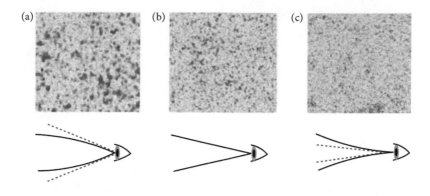

Fig. 18.13 The hot and cold spots appear (a) larger than the actual size if the universe is closed, (b) the same size if the universe is flat, and (c) smaller if the universe is open.

(Credit: NASA)

their actual size if the universe is open and of the same size if it is flat (see Fig. 18.13). According to the best measurements astronomers have made, the universe is flat. The original experiment was done in 2000 and later updated with improved data.

The conclusion that the universe is flat does not tell us about the geometrical structure of the universe. There are a number of possible shapes where space bends and distorts, but the parallel lines remain parallel and the angles of a triangle add to 180 degrees At this point, we cannot determine the shape of the universe as the universe we are able to see (a mere 96 billion light years in diameter) is too small. In order to see the bending of space we may have to know larger pieces of the universe.

In order to see possible shapes of the universe, we again restrict ourselves to a planar (two-dimensional) universe. Visualizing the three-dimensional topology is too complicated.

A planar universe in the shape of a square or a rectangle is the obvious choice for a flat universe. Two parallel lines do not meet and always remain parallel. However, an important feature of the universe is that it is homogeneous, that is, on a big scale, it looks the same in all directions no matter where we are. A planar universe in the shape of a square or a rectangle does not satisfy this requirement. In the middle, the universe will

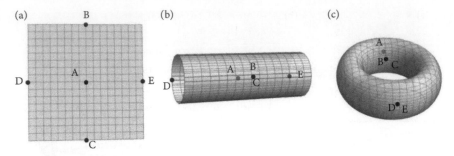

Fig. 18.14 (a) A two-dimensional flat universe in which two parallel lines never meet and the sum of the angles of a triangle is equal to 180 degrees. However, the universe is not homogeneous: for someone at A, the universe looks the same in all directions but not for those at B and C or for those at D and E. (b) If the two-dimensional flat universe is rolled and becomes cylindrical, the symmetry is obtained in the vertical direction. The parallel lines still do not meet and the angles of the triangle add to 180 degrees. (c) The symmetry in the horizontal direction is obtained by rolling the cylinder such that it becomes a doughnut shape.

appear homogeneous but not close to the edges. For example, if we move in the vertical direction as shown in Fig. 18.14a, we will approach the edge of the universe and the universe will not look the same in all directions. What shape should the universe be so that it is flat in the sense that two parallel lines never meet and also fulfills the requirement of homogeneity in the vertical direction? One way is to roll the planar surface in a cylindrical shape as shown in Fig. 18.14b. In a cylindrical shape universe, space is flat and it is homogeneous in the vertical direction. But what about the motion in the horizontal direction? The edge problem is still there. This problem is resolved by bending the cylinder such that the two ends meet each other as shown in Fig. 18.14c. The resulting shape is a doughnut shape.

A doughnut shape is not the only one. Scientists believe that the universe could have one of 18 different shapes.

18.7 Inflationary universe

It is quite remarkable that the big bang theory has been very successful in explaining the evolution of the universe from its birth till the present. As mentioned in Section 18.4, the universe expanded faster than the speed of light in its first moments after the big bang. This era is called the "inflation era."

The word "inflation" is familiar to us in the context of prices of various commodities. When overall prices rise, there is inflation. In cosmology, inflation has a different connotation. Here inflation describes the earliest moments of the universe after the big bang when it expanded at an extremely rapid rate.

How do we know, so confidently, about the "inflation" of the earliest universe? This idea, put forward by Alan Guth and later elaborated by Alexei Starobinsky and Andrei Linde in the 1970s, can resolve three different observations about our present state of universe. It explains why the universe appears to be the same in all directions, why the universe is flat, and why no magnetic monopole exists. The detailed mechanism responsible for the inflation is, however, unknown.

The first problem, usually called the horizon problem, is the uniformity of the observed universe. The universe looks the same when we look in any direction. On a very small scale, there may be a lack of uniformity, but on a bigger scale, it is the same. The strongest evidence for this large-scale uniformity comes from the measurement of the cosmic microwave background radiation, the radiation that we view as the left-over heat from very hot matter in the early universe. The temperature of this left-over radiation is uniform in all directions with an accuracy of one part in a hundred thousand. It is the same temperature in all directions.

A question is: Why is this strange? Why is it a problem?

The problem comes from the origin of the cosmic microwave background. As we discussed, during the first 380,000 years of the universe, this radiation was locked with matter because the matter was in the form of positive and negative charged particles, like electrons and protons, which interact very strongly with radiation. When the universe cooled enough, atoms were formed which were electrically neutral and light could travel in straight lines with almost no interaction with the matter. The situation is like at present when the light from distant stars and galaxies can reach us undisturbed by traveling for billions of light years without being lost.

Since the universe seems to have a uniform temperature at present, the universe must have had this uniform temperature at 380,000 years after the big bang as well. But was it enough time for the universe to have come to a uniform temperature on its own? Things tend to come to a uniform temperature when they are allowed to just sit for some time.

An example can illustrate this point. Let us consider a hot cup of coffee in which a small ice cube is dropped. At first, the ice cube is at a very cold temperature and the coffee surrounding it is hot. As time passes, the ice cube melts, the coffee near the ice cube gets colder but the coffee at a larger distance from the melting ice cube continues to remain hot. There is what we call an inhomogeneous distribution of temperature throughout the coffee. Some parts are colder and some parts are hotter. After a sufficiently long time, maybe 5 minutes, the whole coffee is at a uniform temperature. The final temperature will be slightly lower than the original temperature of the coffee but a lot higher than the temperature of the ice cube.

There are two important points in this example. The first is that it takes a substantial time to reach the same temperature for both the coffee and the ice cube. This time depends on how fast the water molecules of both the coffee and the ice cube are able to travel. The minimum time required to reach a uniform temperature is the time taken by the molecules to travel from the initial location of the ice cube to the most distant points inside the coffee cup. The second point is that, during this time, the temperature distribution is not uniform—some parts of the coffee should be colder than the other parts.

With this example in mind, let us see why a uniform cosmic microwave background is problematic for the usual big bang model.

The problem with the big bang model is that the expansion of the early universe took place at speeds faster than the speed of light. As a result, the matter and the radiation present at that time did not have enough time to settle into an equilibrium state. Here we recall that space can expand at a speed faster than the speed of light but matter and energy cannot as a consequence of the special theory of relativity. The situation is similar to the ice cubes inside the coffee cup. If the coffee cup expands at a rate faster than the speed of the molecules, the coffee will never have the same temperature throughout the cup. The regions around the ice cubes will remain colder than the regions farther away. In a similar manner, two widely separate regions of the observed universe cannot have developed the same temperature because they moved apart from each other faster than the speed of light. It therefore becomes difficult to explain the uniformity of the cosmic microwave background radiation throughout the universe.

The second problem with the usual big bang theory is that it cannot explain why the universe is flat. As discussed in Section 18.6, the density of the universe is very close to the critical density that ensures that the observable universe has no curvature. If this result is projected back in time, the density should be almost perfectly equal to the critical density in the earliest moments after the big bang. Any deviation would be magnified during the process of expansion. For example, if there are extremely small wrinkles on a small globe, they would be magnified if the globe expands to (say) the size of the earth. A simple wrinkle on the globe will become the size of Mount Everest or bigger when expanded. If, within the first few seconds, the universe was just a few percent off from perfect flatness, the universe would either re-collapse or expand so much that it would be almost devoid of matter. How do we explain that the density, and consequently the curvature, is so finely tuned in the expanded universe?

A hypothesis of an ultra-fast expansion, inflation, during the earliest moments (from 10^{-38} to 10^{-36} seconds after the big bang) resolved these issues related to the uniformity and the flatness of the universe.

How can inflation solve the uniformity problem that is usually referred as the horizon problem? In order to understand, we consider the example as depicted in Fig. 18.15. Here we assume that the observed universe at this time is much smaller than the actual universe. Suppose, at the moment of its birth, the universe had some inhomogeneities as represented by two shaded regions. If there was no inflation, these inhomogeneities would have survived during the process of expansion and should be measurable in the observed universe. However, inflation can

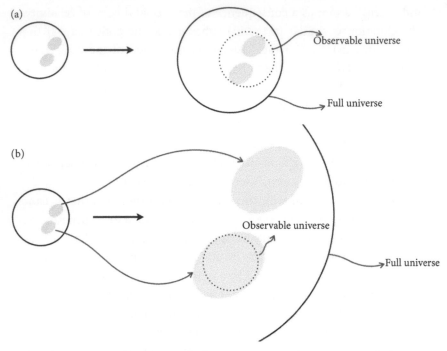

Fig. 18.15 As an illustration, the universe in its earliest moments had some inhomogeneities represented by the shaded regions. The universe expanded faster than the speed of light such that the observable universe becomes smaller in size than the full universe. (a) If the universe had expanded normally, then the observed universe should have the inhomogeneities of the early universe. (b) However, if there was an inflationary expansion in the early moments, then the observed universe is smaller than the scale of the inhomogeneity and would appear homogeneous.

cause an ultra-fast expansion from this initial state. The result is that the observed universe in the expanded universe will contain only one part that is the same throughout. Thus the observed universe, which may be only a small part of the entire universe, will appear homogeneous. This simple example shows how the hypothesis of inflation is able to resolve the horizon problem.

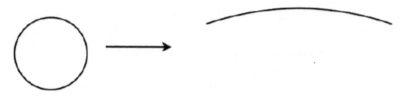

Fig 18.16 The ultra-fast expansion due to inflation flattens the local curvature by making the universe extremely large.

Next, we consider an example to show how cosmic inflation solves the flatness problem as well. A small spherical object like a small balloon has a spherical surface and is strongly curved. However, if the balloon expands to a very large size, the surface over a vast region is almost flat (Figure 18.16). The larger the size of the balloon, the flatter is its surface. Thus the observed universe, which may be a very small part of the entire universe, appears flat.

According to the inflation model the universe expanded 10^{43} (one followed by 43 zeros) times in size during the time between 10^{-36} seconds and 10^{-34} seconds after the big bang. This incredible growth of the universe came to a stop at the end of this period and the universe reverted to its radiation-dominated expansion.

18.8 Unanswered questions

There are many unresolved issues with the big bang model of the universe.

The most obvious question is: What was there before the big bang? Some cosmologists argue that this is a meaningless question as both space and time came into existence at the moment the big bang took place. There is therefore no time before the big bang. But then what does it mean to say that the universe was born 13.8 billion years ago? What was happening (say) 14 billion years ago, 200 million years before the big bang? At this time, this is more like a metaphysical question than a question of science.

Then there is the question: What triggered the big bang? Again, the existing theories are unable to answer this fundamental question.

Another obvious question is that if the universe is expanding, what it is expanding into? When a balloon expands, its two-dimensional surface expands into a third dimension. What about our universe? Is the three-dimensional space expanding into a higher dimension? A typical answer is that the universe is not expanding into an empty space, rather the entire space is expanding. This is something that appears quite incomprehensible.

How is it that a universe born less than 14 billion years ago has the diameter of over a trillion light years when, according to the known laws of physics, nothing moves faster than the speed of light? Well, the answer is that the speed limit applies to the material objects in space. It does not apply to the expansion of space.

How come the universe has three-dimensional space and one-dimensional time? When was it determined that our universe should be four-dimensional: three dimensions for space and one for time?

One of the biggest mysteries is that, based on the movement of the galaxies as well as the expansion rate of the universe, the universe contains much more mass and energy than contained in the observable universe. Almost 95 percent of the universe cannot be accounted for. This huge part of the universe consists of dark matter and dark energy. We turn to the questions relating to these elusive parts of the universe in the next chapter.

19

Dark matter and dark energy

In terms of the most astonishing fact about which we know noth-
ing, there is dark matter and dark energy. We don't know what
either of them is. Everything we know and love about the universe
and all the laws of physics as they apply, apply to four percent of
the universe. That's stunning.

—Neil deGrasse Tyson

When we look up either with the naked eye or with the most sophisticated telescope, an incredibly large universe is in front of us. There is
the whole planetary system consisting of the sun, the planets and their
moons, there are billions of stars forming galaxies, and there are trillions
of galaxies. Besides usual stars, there are other stellar objects like neutron
stars, supernovae, and, of course, black holes. It is a huge and massive
universe. We are tempted to believe that, what we observe is all there is.

One of the most startling and amazing observations about our universe is that what we see or able to see constitutes less than five percent
of the universe. The rest of the universe, over ninety five percent, is hidden from us—it consists of dark matter and dark energy whose effects
we see but cannot see or measure them directly or indirectly. At present
we have no clue about their structure. This is a deep mystery. One of the
most baffling questions facing scientists today is: "What is the nature of
the dark matter that constitutes twenty five percent and the dark energy
that constitutes another seventy percent of the known universe?"

A coherent and detailed picture of the formation of galaxies, clusters
of galaxies, superclusters, and the eventual fate of the universe, cannot
be formed without knowing the nature and properties of the dark matter
and dark energy. The picture of the universe will remain incomplete
without solving the mystery of dark matter and dark energy.

19.1 Dark matter

The discovery of dark matter happened in two phases. First, in 1933, Fritz Zwicky, an American astrophysicist, conjectured the presence of dark matter based on his studies on a cluster of galaxies, called the Coma Cluster. The Coma Cluster is located about 300 million light years from our galaxy and is the home of thousands of galaxies with each galaxy containing hundreds of millions of stars. It is one of the densest known galactic groupings in the universe.

A galaxy is held together by the gravitational attraction among the stars and other massive objects inside the galaxy. The attraction between stars is weak because they are far apart. As an example, the nearest star from the sun is about four light years away. However, if the density of the stars and the amount of mass inside the galaxy is sufficiently large, the galaxy can be held together by the gravitational force. Most stars move around their galaxy. For example, the sun is thought to orbit the Milky Way in about 200 million years. If the mass inside the galaxy, and consequently the gravitational pull, is small then the stars can fly apart and the galaxy can dissipate.

In exactly a similar manner, galaxies are held within a cluster due to the gravitational pull of other galaxies. The galaxies can move inside a cluster. The movements are determined by the presence of other galaxies and their gravitational pull.

Zwicky made a measurement of how fast the galaxies within the Coma Cluster were moving. He found that these galaxies were moving too fast to be held together by the observed matter in the cluster—matter formed by all the galaxies inside the cluster. The only way they could be held together was if there was more mass than what is visible. Zwicky conjectured the presence of "dark matter" that is not visible by the measurement apparatus. About 90 percent of the mass of the Coma Cluster is believed to be in the form of dark matter (Fig. 19.1).

This was a startling discovery. Even more surprising was the lack of interest in this discovery for almost 30 years. During this period, astrophysicists were mostly interested in the expansion of the universe and the big bang theory of the creation of the universe.

Fig. 19.1 Photograph of Coma Cluster. Almost every object in this photo is a galaxy. The Coma Cluster of galaxies contains as many as 10,000 galaxies, each housing billions of stars.

(Credit: NASA, ESA, J. Mack (STScI) and J. Madrid (Australian Telescope National Facility))

In the 1970s, Vera Ruben and Ken Ford were studying the motion of spiral galaxies like our own, the Milky Way (Fig. 19.2). The expectation was that most of the mass in such galaxies is concentrated near the center of the galaxy. Thus gravity is greater near the center and becomes gradually weaker as we move farther away toward the edge of the galaxy. Therefore, stars closer to the center of the galaxy should move faster than the stars at the edge of a galaxy.

What was found was that rotation speed does not diminish but remains roughly the same, all the way to the edge of the galaxy. This was quite surprising. The only way these results could be explained was by assuming that there is more matter in the galaxy than what could be observed. Another piece of evidence for dark matter was found.

This prompted a systematic search for dark matter. A major tool employed to study the dark matter was the gravitational lensing effect that was predicted by Einstein's theory of relativity (Fig. 19.3).

As we discussed in Section 16.3, light from a distant star is bent by the sun's gravitational field due to the bending of the space-time in the

Fig. 19.2 A picture taken by the Hubble Space Telescope of a spiral galaxy. Nearly 70 percent of the galaxies closest to the Milky Way are spiral galaxies.

(Credit: NASA/ESA/Hubble Heritage Team)

vicinity of the massive sun. This was the essence of Eddington's classic experiment to test the predictions of the general theory of relativity. Just as a glass lens can give a focusing effect, a massive object like a galaxy can lead to a similar lensing effect. Here, a galaxy which may contain billions of stars with each star being several light years away from the nearest neighbor, is considered a homogeneous distribution of matter. A galaxy exerts gravitational force on the neighboring galaxies like between two massive objects.

Fig. 19.3 Schematics of a gravitational lens. Light from a bright quasar is bent by the galaxy on its way to earth. The Hubble telescope sees four images, indeed a ring, of the quasar due the bending of light due to the space-time structure around the galaxy, thus forming a gravitational lens. (Credit: NASA)

The gravitational lensing effect can be used to measure the mass distribution of the galaxy or the cluster of galaxies as the bending of the light rays depend upon the enclosed mass. The larger the mass the more the light bends. Visible galaxies are not the only possible gravitational lenses. Dark matter can also reveal its presence by producing this effect. Over the years, gravitational lensing has been used as a tool, not only to calculate the total mass inside a cluster but also how this mass is distributed. The result of such detailed studies has indicated that individual galaxies as well as clusters contain more than ten times as much dark matter as observable matter.

19.2 Search for dark matter

Two questions of great interest are: "What is dark matter?" "What it is composed of?" "Dark" means that it is not visible to us. However, there is the possibility that it may be present in the form of those objects that are not directly visible to us. The most serious candidates are black holes. We discussed in Chapter 17 that black holes cannot be seen directly by any means. Indirect methods are applied to observe them. Maybe there is a large number of black holes where these indirect methods do not work and they are hidden from us. Black holes are not the only invisible objects. There could be white dwarfs whose luminosity is so low that our observational instruments may not be able to detect them. Objects such as black holes and white dwarfs can however act as gravitational lenses as discussed above. They can focus light from a distant star. These objects are dubbed as MACHOs (Massive Compact Halo Objects).

Suppose that an invisible MACHO moves between earth and a star. The star appears to become brighter during the period that the MACHO acts as a gravitational lens. The star returns to its normal brightness when the MACHO has passed. This may take a few hours to a few days. Research teams have observed millions of stars in search of MACHOs. Several have been detected but they are not in such quantity that could explain the dark matter even in our galaxy, the Milky Way.

A serious class of candidates for dark matter since 1985 are the so-called weakly interacting massive particles or WIMPs. These are hypothetical electrically neutral and slowly moving particles. The precise nature of these particles is not known. These particles have been conjectured while extending the existing theories describing the various forces in nature. Despite a long and serious effort, they have not been detected so far. The search still goes on.

Another possibility is that dark matter consists of particles called axions. Axions first made their way into scientific discourse in 1977 to fix some problems with the strong force, the force that keeps atomic nuclei together. They have, however, never been observed. Axions are supposed to be particles with extremely small mass, but the possibility exists that there may be a very large number of them. Since they do not interact with anything else, they may still be there and account for the dark matter.

There are other possibilities as well but, until observed, they remain a matter of speculation. It is amazing that the mysterious substance forming the dark matter that outweighs all the known stars and galaxies in the known universe, has eluded detection so far.

19.3 Dark energy

The big bang, cosmic inflation, the expansion of the universe, the cosmic microwave background radiation—they all seem to explain the history of the universe up to the present time quite successfully. The question is what about the future? It all depends on the matter, visible and dark, that is contained in the universe. The gravitational force leading to the mutual attraction between cosmic objects should be responsible for the future evolution of the universe. This attractive force can slow down the expansion. Would this be sufficient to stop the expansion at some point in time and the beginning of a contraction, eventually collapsing back into a "singularity"? Or would the force be less than what is required for stopping the expansion and the universe continues in expanding mode forever? Or perhaps the forces between the expansion and gravitation may become delicately balanced, leading the universe to come to a standstill? Until the late 1990s, astronomers did not know whether the expansion would continue forever or whether the universe had enough mass to reverse the expansion and revert back to the same kind of "singularity" that initiated the big bang.

In the 1990s, two teams of scientists, the Supernova Cosmology Project led by Saul Perlmutter and the High-z Supernova Search Team led by Brian Schmidt and Nicholas Suntzeff, embarked on stellar observations to answer these questions. Adam Riess also played a crucial role as a member of the High-z Team. Their objective was to study the growth of the universe by analyzing the brightness of some distant objects. These objects included the supernovae produced as a result of exploding stars. The level of brightness is a measure of how far away these objects are as, for the same luminosity, the brighter are nearer than the dimmer ones. In particular, they searched for a specific type of supernova, called supernova 1a, which all explode with about the same energy. The brightness of such supernovae could help astronomers to determine the cosmological distances. If the distance to these bright objects is found to be farther than what is predicted by the big bang model, then the universe must be expanding faster than our original estimate. The luminosity of the supernova can then provide the information about how fast the universe is expanding.

In 1998, the two groups published their results. What they found was that these supernovae, six or seven billion light years away, were not as bright as they expected on the basis of the standard big bang model based only on gravitational forces. This meant that these supernovae were farther away than anticipated. The only explanation for these larger distances is that the universe expanded much more during this period (roughly half the age of the universe) than expected. This led to the conclusion that the expansion rate of the universe is not slowing down. Indeed, it is speeding up. This was a startling result.

This result meant that the dominant factor in the evolution of the universe was not gravitation. There was some "dark energy" that was responsible for this unexplained expansion. A mysterious result was that the expansion was slowing down until seven or eight billion years after the big bang as would be expected due to the gravitational pull between the stellar objects. But then, for some unknown reasons, an anti-gravity force started to dominate, overcoming the brake that the gravitational force was placing on the expansion, reversed the slowing down of the expansion, and started to accelerate (Fig. 19.4). Perlmutter, Riess, and Schmidt were awarded the 2011 Nobel Prize for this discovery.

What is dark energy? Apart from "knowing" that roughly 68 percent of the universe is dark energy that effects the expansion of the universe, there is nothing known about it. It remains one of the biggest mysteries. Any question about the future of the universe cannot be answered in a satisfactory manner without knowing the nature and the properties of the dark matter and dark energy that make up more than 95 percent of the universe. There are many unanswered questions about dark energy.

Scientists have explored many possible explanations to account for dark energy. The problem is that all the explanations have problems.

One explanation is that dark energy is the property of space. When new space is created in the process of the expansion of the universe, the newly generated empty space can possess its own energy. This is an argument that was essentially made by Einstein when he wrote the equations of the general theory of relativity. We recall that a consequence of these equations was that the universe was not static, it could expand or compress depending upon the matter and energy structure. Einstein introduced a cosmological constant to ensure that the universe was not

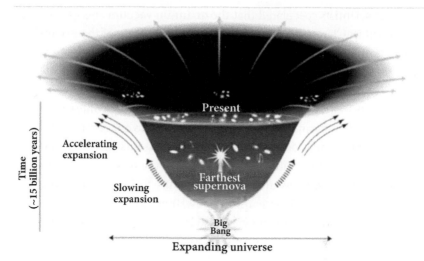

Fig. 19.4 This figure shows that the rate of expansion of the universe increased substantially about 7.5 billion years after the big bang. Astronomers theorize that the faster expansion is due to a mysterious dark force that is pulling galaxies apart.

(Credit: NASA)

expanding. The way this constant ensured a static universe was that a positive or negative energy is generated when the universe expands or shrinks and this energy is the property of space. We also recall that, upon learning of Hubble's observations that the universe was expanding, Einstein abandoned this cosmological constant, calling it the biggest blunder of his life. With the discovery of dark energy, the cosmological constant and its ability to account for missing energy is making a comeback. Einstein might not have been wrong after all. The problem is that no one understands why the cosmological constant should be there. Even more problematic is why it would have exactly the right value for the observed rate of expansion.

Perhaps dark energy is a new kind of energy fluid that fills all space. The nature of this fluid, if it exists, is very peculiar. Its effect on the expansion of the universe is opposite to that of the usual matter and energy.

Some scientists speculated that the quantum vacuum energy, that we discussed in Chapter 9, is responsible for the increase of the expansion rate. This idea is not new. Many years before the enhanced cosmic expansion was discovered, it was realized that empty space is filled with a large amount of energy associated with the field fluctuations in vacuum. This vacuum energy, when converted into mass density, comes out to be equal to 10^{93} grams in a cube with each side equal to one centimeter. This is an extremely high density. According to Einstein's theory of relativity, the space-time curvature is determined by the mass-energy content. Thus, vacuum energy would cause a curvature too large. Also, such an energy would lead to an expansion. How, then, to resolve this discrepancy? A resolution was sought via Einstein's cosmological constant. Einstein assigned the value of this constant to be zero after Hubble's observation that the universe was expanding. Physicists now resort to a non-zero value of the cosmological constant. Its value was chosen such that it balanced the vacuum energy density, thus no extra expansion would occur as a consequence of such a large vacuum energy. What would cause such a large balancing energy remained an unsolved problem.

When evidence for dark energy appeared, it was thought that this could explain the balancing energy needed to cancel the vacuum energy. However, soon it was realized that the density of the dark energy was incredibly small, only 10^{-27} gram in a cube with each side equal to one centimeter. This difference, 10^{93} versus 10^{-27}, is one of the largest differences for an explanation. The vacuum energy being the dark energy was thus ruled out. But the mystery remains about how to explain the gravitational effect of such a large vacuum energy.

The most drastic explanation casts doubt on the correctness of Einstein's general theory of relativity. A modification may be able to solve this cosmic riddle. The problem is that this theory has passed all the tests on the smaller cosmic scale like the perihelion of Mercury and the bending of light by the sun. A modified theory will have to correctly predict the observed expansion on a scale corresponding to almost half the age of the universe while yielding the right results for the stellar objects in the solar system. This appears to be a tall order.

In the absence of a satisfactory explanation, the mystery, perhaps the greatest in the history of mankind, continues.

PART 4

EPILOGUE

20

Dreams of the future

How has this come about? Somehow the universe has engineered, not just its own awareness, but also its own comprehension. Mindless, blundering atoms have conspired to make not just life, not just mind, but understanding.

—*Paul Davies*

After studying the laws of nature as they stand, questions still remain. Why is the universe like this? What determined the laws that govern the universe and who determined them?

Some other questions also arise. Do we expect to see a major upheaval of the kind that happened in the early years of the twentieth century? Will the laws of nature as understood today go through a major revision at some point in the future?

We are living in an era when technology is moving at a dizzying pace and new levels of precision in our measurement devices are bringing new perspectives concerning the mysterious laws that govern our universe. However, the remarkable result is that, after the passage of more than a hundred years, there is no observation that is at variance with the predictions of quantum mechanics and the theory of relativity.

This does not mean that science has come to an end and no issues remain unresolved. Far from it. Sometimes it seems like we have touched only the tip of the iceberg and some of the major fundamental questions stay unresolved. A list of such issues may include the following.

- Can gravitation be combined together with the other forces of nature in a single unified force?
- Is there only one universe or there are multiverses?
- Why is our universe just right for life?

- Can we understand what life is? Is it something more than a collection of atoms and molecules?
- What is the nature of dark matter and dark energy?
- Humans have free will. Can this free will be incorporated into the laws of nature?
- Will we be able to restore full determinism fulfilling Einstein's dream of a "complete" theory?

The list of unanswered questions can go on. In this concluding chapter, the significance of some of these questions is presented.

20.1 Quantum gravity

Quantum mechanics and the theory of relativity form the foundation of the laws that govern our universe. However, so far these theories have maintained their independent existence. A big challenge is to unify these two highly successful theories into a single unified theory. This problem is not merely a theoretical problem—a unification is necessary to fully comprehend the earliest phase of the universe. This would have consequences on answering the questions like: Why is our universe like what it is? And how did the laws that govern the universe came about? A theory of quantum gravity would also shed light on the nature of black holes.

The universe, as we understand it, is held together by four forces. These are electromagnetic, weak nuclear, strong nuclear, and gravitational forces.

We discussed in Chapter 5 how the three forces of nature, the electromagnetic, the weak, and the strong forces, can be described quantum mechanically. During the 1970s and 1980s, these three forces were combined together into a Grand Unified Theory (usually referred as GUT). It remains an unfulfilled dream to bring gravity within the fold and create a Theory of Everything. Such a theory will be able to explain the working of the cosmos from first principles.

We recall (from Section 18.4) that all the four forces were united at the moment the big bang took place. At that moment, the temperature was extremely high. When the universe cooled a bit after 10^{-43} second, the gravitational force separated from the other three forces. Gravitational force, however, played a key role in the further evolution of the universe, forming stars and galaxies.

There is a fundamental difference between gravitation and the rest of the forces. The electromagnetic, the weak, and the strong forces act between massive as well as massless objects. These forces are based on exchange of quanta of energy that can behave both like a particle and like a wave. As an example, electromagnetic force takes place via an exchange of photons. However, gravitation comes about due to the curvature of space and time according to the general theory of relativity. Gravity is not a force acting within space and time like the other three, it is linked to the fabric of space and time itself. A theory of quantum gravity would involve quantizing space and time.

A major obstacle in unifying quantum mechanics and gravitation is that gravitational effects are immensely weak as compared to the other forces. Gravitation plays a key role in the movement of huge and massive objects like planets, stars, and galaxies. At the atomic level, where quantum effects are strong, gravity plays essentially no role.

A theory of quantum gravity becomes important in regions of the size smaller than the Planck length which is equal to the incredibly small size of 10^{-35} meter. At such a size, the energy density, and consequently the temperature, is so high that the laws of classical mechanics cease to exist and a quantum mechanical description is required. A region of such size, for example, existed at the moment the big bang took place. In order to understand what caused the big bang, it is necessary to understand what happened during those initial moments when the universe was the size of Planck's length.

In a similar manner, a theory of quantum gravity is needed to fully understand black holes when the size of the black hole is less than Planck's length.

A way to see that gravitation is a quantum effect is to explore the quantum nature of the gravitational waves that arise due to a changing distribution of massive objects. For example, do these gravitational waves have the kind of quantum fluctuations that we associate with electromagnetic waves? Do these have particle characteristics, a graviton, like the electromagnetic waves can be described in terms of photons?

These are still open questions mainly due to the very weak nature of gravitational waves.

20.2 What is life?

Another question that has been addressed by scientists, philosophers, and theologians relates to the origin and nature of life. Even with the stunning success of scientific theories in explaining the laws of nature, no one seems to know where life comes from. Is life something external or the result of certain chemical processes?

One question is whether the smallest living object, such as a cell, can be replicated atom by atom. Would the resulting object acquire all the traits of the cell? No one seems to know the answer to this question at the present time. However, the answer may come sooner than we think. We are approaching a point in our scientific progress where we can isolate and address single atoms and fabricate a device atom by atom.

An important milestone will come if it becomes possible for scientists to make an identical copy of an intelligent living object—a human being—atom by atom. This may be a far-fetched dream at this time. However, it should be possible to build such a human being at some point in the future. The question is whether such a human being will become alive with features like free will, a thinking mind, a sense of good and bad, and reproductive capability. Would such an object have emotions? One of the most important questions is whether life is something beyond a particular arrangement of atoms and molecules. Is there a soul that transcends our physical composition?

This question is particularly important from a theological point of view. In most theological doctrines or religions, a soul is considered eternal. The body is just a vessel to carry the soul—the body may die and decay but the soul lives forever. This concept lies at the foundation of practically all religions. An answer to the question whether life and soul are independent of our body and are eternal would have decisive consequences on the validity of religious doctrines.

A related question is whether we can reach a point where the human dream of an eternal life in this world is realized. Defeating death would be the capstone of scientific achievement. What has become possible in our lifetime is that we have an understanding of the basic components of life such as a cell. In some ways, a human body is perceived like a machine consisting of a large number of nuts and bolts or organs. When some of these organs fail, like a part of a machine, we can fix them or replace them. Transplanting important organs like heart and kidney has become almost routine. But death is considered irreversible. This is a big mystery. When someone is proclaimed dead, there is no hope to bring him back to life. Will it be possible some day to use physics and biotechnology to fix the human "machine" like, given resources, a car can be fixed no matter how damaged it is? Will scientific progress help conquer death?

Then the question is whether science as understood today allows these possibilities. Is there something fundamental in the laws of physics that does not allow replicating a human object or bringing a dead person back to this world again?

20.3 Free will and the laws of nature

There is another important issue. Human life is endowed with a free will. An alive creature like a human being has the capability to affect the laws of nature. For example, according to the law of gravitation, if a ball is dropped, it should hit the ground and then keep bouncing. But someone can decide to catch the ball after just one bounce, thus interrupting the laws of nature by invoking her free will. Free will has the power to affect or change the laws that govern this universe. This raises the important question: What is free will? Can it be quantified and included in the existing physical laws?

As another example, suppose a creature in another galaxy is watching our galaxy and is somehow able to see the motion of the planets around the sun. Knowing the periodicity of the planets' motion and the distances between these stellar objects, they will conclude that everything is according to the laws of nature, in this case, the law of gravitation. All this is based on the assumption that the sun and the planets are nonliving objects.

Next, suppose some intelligent humans are able to break a certain stellar object like the moon into pieces. This can, at least in principle, be done by sending a big rocket full of a large number of atomic and hydrogen bombs that are exploded on the surface of the moon. This action involves human interference and cannot be considered a natural phenomenon. The splitting of the moon will change the motion of the planets including earth in a perceptible way. The creatures far away, unaware that an intelligent creature with free will lives on the planet earth, will be baffled by the modified motion of the planetary system. To them, the planetary system would not appear to be following the Newtonian laws of motion.

The only thing that is added in this scenario to the existing laws that govern the dynamics of nonliving objects is the free will of humans. Again, this raises an important question: Is human life and its associated free will something beyond a collection of atoms and molecules in a certain order? How do we include free will within the laws of nature?

20.4 Why is the universe just right for life?

One of the major mysteries concerning the laws of nature is that the universe is well suited for the emergence of life as we know it. The universe appears to be delicately fine-tuned for our existence. Here is a partial list of the coincidences that allowed life forms like ours to exist and ponder over the laws that so beautifully determine the evolution of this universe.

- The ratio of the strength of the electromagnetic force and the gravitational force between two protons is about 10^{36} (1 followed by 36 zeros). If the ratio was only slightly different, the universe would have collapsed before stars and galaxies were formed.
- If the ratio of the masses of the neutron and proton (1.00137841931) is flipped, making the proton heavier than the neutron, atoms would not have been possible. There could not have been the universe and life as we see it.
- As discussed in Chapter 18, there are two competing forces in the universe—one is responsible for its expansion and the other is

gravity that is trying to stop it. This balance is responsible for the existence of the universe in its present form. According to Stephen Hawking, "If the rate of expansion one second after the big bang had been smaller by even one part in 100 thousand million million, the universe would have collapsed before it ever reached its present size."

- If the electric charge of the electron had been only slightly different, stars would not have been able to burn hydrogen and helium.
- If the density of the universe one second after the big bang had been greater by one part in a thousand billion, the universe would have collapsed after 10 years. On the other hand, if the density of the universe at that time had been less by the same amount, the universe would have been essentially empty with no stars and galaxies since it was about 10 years old.

How is it that the universe is so fine-tuned to be what it is? In particular, the suitability of the universe for a life form like ourselves makes us wonder whether it is a fortunate accident or it was designed this way. After all, without us, there would not have been conscious observers in the universe. In order to remove any special status to our universe, many cosmologists believe that our universe is just one of an almost infinite number of universes. The parameters may be all very different in all those universes and it is a happy coincidence that we are in the universe that has just the right values of the parameters for life to exist. It is unclear whether an answer to the question concerning the presence of just one universe or multiverses will ever come.

Bibliography

Chapter 1

J. Al-Khalili, *The World According to Physics* (Princeton University Press 2020).

R. J. Scully and M. O. Scully, *The Demon and the Quantum* (John-Wiley-VCH 2010).

S. Weinberg, *To Explain the World: The Discovery of Modern Science* (HarperCollins Publisher, New York 2015).

M. S. Zubairy, *Quantum Mechanics for Beginners with Applications to Quantum Communication and Quantum Computing* (Oxford University Press 2020).

Chapter 2

R. P. Feynman, *Six Easy Pieces: Essentials of Physics Explained by its Most Brilliant Teacher* (Basic Books 2011).

R. P. Feynman, R. Leighton, and M. Sands, *The Feynman Lectures on Physics, Vol. I* (Addison-Wesley, Reading, MA 1965).

D. C. Giancoli, *Physics: Principles with Applications* (Pearson 2013).

L. Susskind and G. Hrabovsky, *The Theoretical Minimum: What you Need to Know to Start Doing Physics* (Basic Books 2013).

H. D. Young and R. A. Freedman, *University Physics* (Pearson 2015).

Chapter 3

J. Baggott, *The Quantum Story: A History in 40 Moments* (Oxford University Press 2011).

L. de Broglie, *The wave nature of the electron*, Nobel Lecture (1929).

D. M. Greenberger, N. Erez, M. O. Scully, A. A. Svidzinsky, and M. S. Zubairy, *The rich interface between optical and quantum statistical physics: Planck, photon statistics, and Bose–Einstein condensates*, in Progress in Optics, Vol. 50, Edited by E. Wolf (Elsevier, Amsterdam 2007), p. 275.

H. F. Hameka, *Quantum Mechanics: A Conceptual Approach* (John Wiley 2004).

K. A. Peacock, *The Quantum Revolution: A Historical Perspective* (Greenwood Press 2008).

J. Polkinghorne, *Quantum Theory: A Very Short Introduction* (Oxford University Press 2002).

M. G. Raymer, *Quantum Physics: What Everyone Needs to Know* (Oxford University Press 2017).

Chapter 4

O. Darrigol, *A History of Light: From Greek Antiquity to the Nineteenth Century* (Oxford University Press 2012).

A. M. Smith, *From Sight to Light: The Passage from Ancient to Modern Optics* (University of Chicago Press 2015).

M. S. Zubairy, *A very brief history of light*, in Optics in Our Time, Edited by M. D. Alamri, M. M. El-Gomati, and M. S. Zubairy (Springer Nature 2016).

Chapter 5

M. Born, *Atomic Physics* (Dover Publications 1989).

J. Challoner, *The Atom: A Visual Tour* (MIT Press 2018).

R. P. Feynman, *QED: The Strange Theory of Light and Matter* (Princeton University Press 2014).

H. Kragh, *Niels Bohr and the Quantum Atom: The Bohr Model of Atomic Structure 1913–1925* (Oxford University Press 2012).

B. Pullman, *The Atom in the History of Human Thought* (Oxford University Press 2001).

Chapter 6

C. H. Bennett and G. Brassard, *Quantum cryptography: Public key distribution and coin tossing*, in Proceedings of IEEE International Conference on Computers, Systems, and Signal Processing, Bangalore, India (1984), pp 175–179.

C. H. Bennett, G. Brassard, and A. K. Eckert, Quantum cryptography, *Scientific American* **267**, 50 (1992).

C. H. Holbrow, E. Galvez, and M. E. Parks, Photon quantum mechanics and beam splitters, *American Journal of Physics* **70**, 260 (2002).

P. Lambropoulos and D. Petrosyan, *Fundamentals of Quantum Optics and Quantum Information* (Springer 2007).

A. Rae, *Quantum Physics: A Beginner's Guide* (Oneworld 2005).

V. Scarani, L. Chua, and S. Y. Liu, *Six Quantum Pieces: A First Course in Quantum Mechanics* (World Scientific 2010).

Chapter 7

J. A. Barrett, *The Conceptual Foundations of Quantum Mechanics* (Oxford University Press 2019).

S. Carroll, *Something Deeply Hidden: Quantum Worlds and the Emergence of Space-time* (Dutton 2020).

A. Pais, *Niels Bohr's Times, in Physics, Philosophy, and Polity* (Oxford University Press 1991).

Chapter 8

H. C. Corben, Another look through the Heisenberg microscope, *American Journal of Physics* **47**, 1036 (1979).

B. G. Williams, Compton scattering and Heisenberg's microscope, *American Journal of Physics* **52**, 425 (1984).

W. K. Wooters and W. H. Zurek, A single photon cannot be cloned, *Nature* **299**, 802 (1982).

Chapter 9

C. Q. Choi, Something from nothing? A vacuum can yield flashes of light, *Scientific American* (February 2013).

P. W. Milonni, The *Quantum Vacuum: An Introduction* to *Quantum Electrodynamics.* (Academic Press 1994).

J. Munday, A new twist on the quantum vacuum, *Physics Today* **72**, 10, 74 (2019).

Chapter 10

Y. Aharonov and M. S. Zubairy, Time and the quantum: Erasing the past and impacting the future, *Science* **307**, 875 (2005).

R. P. Feynman, R. Leighton, and M. Sands, *The Feynman Lectures on Physics, Vol. III* (Addison-Wesley, Reading, MA 1965).

B. Greene, *The Fabric of the Cosmos* (Alfred A. Knopf, New York 2004).

C. Jönsson, *Zeitschrift für Physik*, **161**, 454 (1961); translated in C. Jönsson, Electron Diffraction at Multiple Slits, *American Journal of Physics* **42**, 4 (1974).

Y.-H. Kim, R. Yu, S. P. Kulik, Y. Shih, and M. O. Scully, Delayed "choice" quantum eraser, *Physical Review Letters* **84**, 1 (2000).

M. O. Scully and K. Drühl, Quantum eraser: A proposed photon correlation experiment concerning observation and "delayed choice" in quantum mechanics, *Physical Review A* **25**, 2208 (1982).

M. O. Scully, B.-G. Englert, and H. Walther, Quantum optical tests of complementarity, *Nature* **351**, 111 (1991).

R. J. Scully and M. O. Scully, *The Demon and the Quantum* (John-Wiley-VCH 2010).

M. O. Scully and M. S. Zubairy, *Quantum Optics* (Cambridge University Press 1997).

S. P. Walborn, M. O. Terra Cunha, S. Pádua, and C. H. Monken, Double-slit quantum eraser, *Physical Review A* **65**, 033818 (2002).

J. A. Wheeler, The 'Past' and the 'Delayed-Choice Double-Slit Experiment', in A.R. Marlow, editor, *Mathematical Foundations of Quantum Theory*, Academic Press (1978).

Chapter 11

S. Barnett, *Quantum Informatics* (Oxford University Press 2009).

C. H. Bennett, G. Brassard, C. Crépeau, R. Jozsa, A. Peres, and W. K. Wootters, Teleporting an unknown quantum state via dual classical and Einstein–Podolsky–Rosen channels, *Physical Review Letters* **70**, 1895 (1993).

E. Farhi and S. Gutmann, Analog analogue of a digital quantum computation, *Physical Review A* **57**, 2403 (1998).

J. Gribbin, *In Search of Schrödinger's Cat: Quantum Physics and Reality* (Bantam 1984).

L. K. Grover, Quantum mechanics helps in searching for a needle in a haystack, *Physical Review Letters* **79**, 325 (1997).

A. Muthukrishnan, M. Jones, M. O. Scully, and M. S. Zubairy, Quantum shell game: finding the hidden pea in a single attempt, *Journal of Modern Optics* **16**, 2351 (2004).

M. O. Scully and M. S. Zubairy, Quantum optical implementation of Grover's algorithm, *Proceedings of the National Academy of Sciences (USA)* **98**, 9490 (2001).

P. Shor, in Proceedings of the 35th Annual Symposium on Foundations of Computer Science, Santa Fe, NM, Edited by S. Goldwasser, (IEEE Computer Society Press, New York, 1994), p. 124.

J. A. Wheeler and W.H. Zurek (eds.), *Quantum Theory and Measurement* (Princeton University Press 1983).

Chapter 12

A. Aspect, J. Dalibard, and G. Roger, Experimental test of Bell's inequalities using time-varying analyzers, *Physical Review Letters* **49**, 1804 (1982).

J. Bell, On the Einstein–Podolsky–Rosen paradox, *Physics* **1**, 195 (1965).

N. Bohr, Can quantum-mechanical description of physical reality be considered complete?, *Physical Review* **48**, 696 (1935).

A. Einstein, B. Podolsky, and N. Rosen, Can quantum-mechanical description of physical reality be considered complete?, *Physical Review* **47**, 777 (1935).

S. J. Freedman and J. F. Clauser, Experimental test of local hidden-variable theories, *Physical Review Letters* **28**, 938 (1972).

E. S. Fry and R. C. Thompson, Experimental test of local hidden-variable theories, *Physical Review Letters* **37**, 465 (1976).

Chapter 13

A.J. Leggett, The quantum measurement problem, *Science* **307**, 871 (2005).

A. Rae, *Quantum Physics: Illusion or Reality?* (Cambridge University Press 2012).

B. Rosenblum and F. Kuttner, *Quantum Enigma* (Oxford University Press 2011).

H. P. Stapp, The Copenhagen interpretation, *American Journal of Physics* **40**, 1098 (1972).

Chapter 14

A. Fazekas and H. Schneider, *National Geographic Backyard Guide to the Night Sky* (National Geographic 2019).

L. Harvey-Smith, *The Secret Life of Stars* (Thames & Hudson 2021).

S. W. Stahler, The inner life of star clusters, *Scientific American* **308**, 44 (2013).

Chapter 15

S. M. Carroll, The Biggest Ideas in the Universe: Space, Time, and Motion (Dutton 2022).

A. Einstein, *Relativity: The Special and General Theory* (Dover 2010).

J. H. Smith, *Special Relativity* (Dover 2015).

Chapter 16

R. R. Gould, Why does a ball fall?: A new visualization for Einstein's model of gravity, *American Journal of Physics* **84**, 396 (2016).

J. Higbie, Gravitational lens, *American Journal of Physics* **49**, 652 (1981).

A. Pais, *Subtle Is the Lord....The Science and the Life of Albert Einstein* (Oxford University Press 1982).

R. H. Price, Spatial curvature, spacetime curvature, and gravity, *American Journal of Physics* **84**, 588 (2016).

C. Rovelli, *General Relativity: The Essentials* (Cambridge University Press 2021).

Chapter 17

J. Al-Khalili, *Black Holes, Wormholes and Time Machines* (CRC 2011).

B. Cox and J. Forshaw, *Black Holes: The Key to Understanding the Universe* (Mariner Books 2023).

S. Hawking, *Black Holes: The Reith Lectures* (Penguin Random House 2001).

K. Thorne, *Black Holes and Time Warps* (W. W. Norton 1995).

D. Toback, *Big Bang, Black Holes, No Math* (Kendall Hunt Publishing 2013).

Chapter 18

N. de Grasse Tyson, *Astrophysics for People in a Hurry* (W. W. Norton 2017).

A. Delsemme, *Our Cosmic Origins: From the Big Bang to the Emergence of Life and Intelligence*, (Cambridge University Press 1998).

H. Satz, *Before Time Began: The Big Bang and the Emerging Universe* (Oxford University Press 2017).

S. Weinberg, *The First Three Minutes: A Modern View of the Origin of the Universe* (Basic Books 1993).

Chapter 19

B. Clegg, *Dark Matter and Dark Energy* (Icon Books, 2019).

S. Hossenfelder and S. S. McGaugh, Is Dark Matter Real?, *Scientific American* **319**, 2, 36–43 (August 2018).

R. Panek, *The 4 Percent Universe: Dark Matter, Dark Energy, and the Race to Discover the Rest of Reality* (Mariner Books 2011).

L. Randall, What Is Dark Matter?, *Scientific American* **318**, 6, 58–59 (June 2018).

Chapter 20

D. Atkatz, Quantum cosmology for pedestrians, *American Journal of Physics* **62**, 619 (1994).

P. Davies, *A Goldilocks Enigma: Why Is the Universe Just Right for Life?* (Mariner Books 2008).

J. R. Gott, *Time Travel in Einstein's Universe* (Mariner Books 2002).

M. Kaku, *Parallel Worlds: A Journey through Creation, Higher Dimensions, and the Future of the Cosmos* (Anchor 2006).

R. Penrose, *Shadows of the Mind: A Search for the Missing Science of Consciousness* (Oxford University Press 1996).

Index

For the benefit of digital users, indexed terms that span two pages (e.g., 52–53) may, on occasion, appear on only one of those pages.